FOUNDATIONS
of NATURAL
FARMING

UNDERSTANDING CORE CONCEPTS
OF ECOLOGICAL AGRICULTURE

FOUNDATIONS *of* NATURAL FARMING

UNDERSTANDING CORE CONCEPTS OF ECOLOGICAL AGRICULTURE

Harold Willis, Ph.D.

Acres U.S.A.
Austin, Texas

Foundations of Natural Farming

Copyright © 2008 Harold L. Willis

Acres U.S.A.
P.O. Box 91299
Austin, Texas 78709 U.S.A.
(512) 892-4400 • fax (512) 892-4448
info@acresusa.com • www.acresusa.com

Printed in the United States of America

Publisher's Cataloging-in-Publication

Willis, Harold L., 1940-
Farming with nature / Harold L. Willis. Austin, TX, ACRES U.S.A., 2008
    x, 374 pp., 23 cm.
    Includes Index
    Includes Bibliography
    Incudes Illustrations
    ISBN 978-1-60173-007-7 (trade)

    1. Organic farming. 2. Alternative agriculture. 3. Sustainable agriculture.
    4. Agricultural ecology. 5. Organic fertilizers. I. Willis, Harold L., 1940-
    II. Title.

    S605.5.W51 2008   631.584

Contents

Foreword

I t is said that some ancient Chinese rulers would give their enemy a sarcastic "blessing," which they intended as a curse. It was, "May you live in interesting times."

As we plod into the brave new world of the 21st century, few would argue that these are interesting times. Times of head-spinning technological developments like faster computers and curing cancer. Times of incredible wealth accumulation . . . for a few . . . but grinding poverty for so many. And times of alarming dangers—polarized political and religious groups that lead to terrorism and ineffective governments, or melting glaciers and rising sea levels from global warming, and oceans nearly empty of fish, with a "dead zone" the size of New Jersey every summer at the mouth of the Mississippi River.

Are things really going the way we would like? Our "advanced" society has given us obese people dying of heart disease and stroke. Gigantic factory farms pour forth massive quantities of eggs, milk and meat that end up in super-duper burgers. Yet burger-eating customers may partake uneasily for fear of *E. coli*. Pregnant women are warned not to eat fish caught in our streams because of mercury pollution.

In developed and developing countries alike, nearly every aspect of life is controlled or influenced by big business—from the way farmers grow crops and the prices they receive for them, to the nutritional content of the processed cereal we feed our children, to the country in which our shoes and shirts were made, to the price we have to pay for the prescription drugs we think we need.

Many of us seem to be living better than our parents and grandparents did . . . yet are we really happy? The thinking person probably wakes up in the morning with a distinctly uneasy feeling and has trouble getting to sleep at night as well. Those who seriously keep up with the news and get complete, accurate news from unbiased sources are well aware that **things just aren't good . . . and they are getting worse!**

The modern political-economic-industrial system is broken, in dire need of change. Not just a few band-aids to patch up the worst spots, but total reworking and renewal. Everything is topsy-turvy. The natural resources and life-sustaining ecological systems of the earth are being pillaged and wasted at increasing pace. Those who tend and nurture the food and fiber crops and animals—those involved in agriculture—are relegated to second-class lives, endless hard work and too-small rewards. As are most parts of modern life, **agriculture is in crisis!**

Agriculture has undergone considerable change since World War II. Yields have increased. So have fertilizer and pesticide use, but too frequently weeds and pests have increased while crop quality and animal health have gone downhill. Average farm size has increased, while the number of farms and farmers has shrunk. Tractors are bigger, but the soil is harder.

What is wrong? Are you really being told the best way to farm?

Consider this. Farmers and ranchers are working with plants and animals—living creatures. All of the marvelous variety of plants and animals, birds and butterflies on the earth function best under certain conditions and operate according to the laws of biology, chemistry and physics. Any living organism—or any cell from any organism—is infinitely more complex than any man-made machine or computer.

Throw a wrench into a jet engine or hit a watch with a hammer and the results are not good. *Is it any wonder that agriculture today is having problems when so many farmers are trying to raise crops and animals while violating some of the biological laws of healthy growth and nutrition?*

Sure, you can get crops and animals to survive and give you some production, and maybe even record-breaking yields . . . but are they at peak health, and are the commodities they produce of top quality?

The emphasis today is almost always on quantity, *not quality.* Both crops and animals are over-fertilized, gorged, activated, supplemented, stimulated and prodded—to produce *more* . . . and MORE!

A growing number of farmers are getting off this not-so-merry-go-round. They are learning how to beat the economic system that is rigged against them. They are learning that they can often reduce their expenses yet grow excellent yields of high quality crops and raise healthy, productive animals which they can market profitably.

How is this possible? There is no magic formula, no secret product that will solve all problems, just common-sense farming. True, you may have to change the way you do things. You may have to try something new (but has the old way given you the results you want?). You will need to treat your soil, crops and animals in ways that are in harmony with the natural laws that promote good health and high quality products. Sound like a dream? It *can* be done, and it IS being done by an increasing number of farmers.

You will need to develop a farming system based on nature's ecological principles. It is called by various names—ecological agriculture, eco-agriculture, biological agriculture, and so on. You don't have to go as far as being certified organic, or following the detailed methods of bio-dynamic agriculture. You can adapt a mixture of natural and technologically advanced methods. The main idea is to work out a farming system that works for you—for your geographic area, soil and climate, crops and animals raised, available markets, etc.

You may need to learn some new things and even un-learn some others that aren't working. You will probably want to talk to other farmers who are having success, go to field days, attend educational conferences, or seek the help of a consultant. Change is difficult, especially when it involves your livelihood. It takes courage, determination and hard work. But the results are worth it. Farming can be enjoyable, satisfying and profitable.

This book is intended to provide a basic foundation of knowledge and ideas to get you started. It will introduce you to the ecological laws of nature that regulate how the soil, plants and animals function. It will introduce you to the basics of soil tillage and fertilization, crop growth, weed and pest problems, and the need for quality. We cannot cover all the nuts-and-bolts details of exactly which equipment you need, how deep to plant, when or whether to use a chemical control, and so on. That is what you will need to work out yourself. But before you do that, *you need to know where you are headed.* You need to know *why* you are doing what you are doing. If there is no real need to use herbicide, *don't do it!* You need to learn to farm smarter in order to survive at farming. So that is what this book is intended to provide—the *foundations* of ecological agriculture.

Chapter 1

Fiddler on the Roof —

TRADITION!

T o those who have seen the wonderful play or movie, *Fiddler on the Roof,* one of the memorable songs is "Tradition!" How often we follow tradition, doing things the way our fathers did or the way our neighbors do, because we are a little bit afraid to break away from the crowd, to try something different. And of course, we don't have confidence in our own knowledge and logic, so we follow the experts' advice—after all . . . they should know, shouldn't they?

Unfortunately, most school children do not particularly like the study of history, and therefore most of us grow up lacking an appreciation of the past or a desire to investigate it. We are so caught up in the whirl of this scientifically, technologically marvelous 21st century that we think (or are told) that only the most up-to-date methods and products are worthwhile. We have forgotten, or never bothered to find out, how our grandfathers used to farm.

What is considered "traditional" or conventional agriculture today is hardly old enough to be dry behind the ears, much less be truly traditional. It is mostly a post-World War II phenomenon. Before we are finished with our journey, we may find that the *real* traditional agriculture, the way our grandfathers farmed, has more value than the bells and whistles of "modern" agriculture.

One of the few advantages of growing old is that you can remember what it was like in the "good old days." If you are old enough, or just sit down and think for a moment, or better yet, talk to an old-timer, you will realize that *there are more problems now than there used to be.* Worse insect problems. Worse weed problems. Corn blight and moldy corn. Cattle with mastitis and breeding failure. Even soil is harder, requiring larger tractors to till it. Most of these troubles have appeared, or at least have become widespread, since World War II. Could there be a connection between this and the fact that modern farming methods have also become widespread—"traditional"—since World War II?

The origins of modern agriculture

What is "modern agriculture"? Is it the gigantic tractors and plows pictured in the glossy pages of the farm magazines? The latest bug- and weed-zapping chemicals proclaimed on prime-time TV? Is it no-till farming, in the name of soil conservation? Is it 200 bushel per acre corn and supercows that produce 20,000 pounds of milk per year? Yes, it is all that and more. To read the advertisements, plus the optimistic predictions of what science and technology will do for us tomorrow, you would think that we have nothing to worry about. Is everything coming up roses? Or don't roses also have a lot of thorns? Let's look at modern agriculture, how it developed, how it operates, and where it is headed.

What is it? What we call modern agriculture is the methods and systems that are an outgrowth of various scientific and technological developments which have become widespread since World War II. Among the manifestations of modern agriculture are (1) large-scale use of inorganic salt fertilizers, herbicides and other pesticides, and growth stimulants and drugs (for animal agriculture); (2) increased size of farms and fields, with large machinery, the clearing of windbreaks, irrigation, and concentrated animal feeding operations; and (3) improvement of plant and animal characteristics by hybrids, strain improvement, artificial breeding, and most recently, genetic engineering, the direct manipulation of a plant's or animal's genes.

These technological advances began to become widespread from 1941–1945, and have exploded since then into what historians call the Second American Agricultural Revolution.* They have truly spawned a revolution—they have transformed agriculture into a "whole new ball game."

However, these technological developments did not spring into being full blown during World War II. They had their predecessors; they were

*The First American Agricultural Revolution began during the Civil War (1861–1865), when increased demand for food led to the commercialization of agriculture in the northern states.

developed gradually over the decades before. Let's look at how some of them originated.

Fertilizers. Knowledge that adding certain natural materials to the soil (such as lime, gypsum and manure) improves yields has been around since ancient Roman times and before. The first European settlers in Jamestown in 1609 and the pilgrims in 1621 learned from the Indians how to grow corn (maize) by adding a dead fish to each hill. The impetus that gave rise to the modern synthetic fertilizer industry came from the research of the German chemist, Justus von Liebig. After studying plant growth in relation to the amount of materials used up from the soil, he concluded that certain amounts of minerals were needed to produce a certain amount of growth or yield and that a total diet of inorganic minerals could supply the plant's needs. This gave rise to the current "balance-sheet" theory of plant nutrition in which fertilizer amounts are recommended according to the farmer's yield goal, with little or no attention being paid to the contribution of organic matter, nitrogen-fixation, and other beneficial effects of soil organisms.

In the U.S.A., mixed chemical fertilizers first became commercially available in 1849, but the use of lime and synthetic fertilizers simply did not become widespread until World War II, probably partly because farmers previously farmed in harmony with natural systems, but also because World War II brought an incentive to greatly increase yields. More food was needed to help make up for lost production in Europe, and the prices farmers received doubled. The annual use of "N-P-K" fertilizers increased 95% in 1945 over the 1935–39 period, while lime usage increased three times.

Since then, the use of synthetic fertilizers and crop yields have both climbed astronomically. Commercial fertilizer used in 1980 amounted to over 75 million tons, compared to slightly less than 21 million tons in 1950 and only 8.55 million tons in 1940. Now synthetic nitrogen fertilizers are used on 98% of U.S. corn land. U.S. fertilizer use has leveled off since the mid-1970s.

Pesticides. Just as with fertilizers, the use of relatively safe insecticides and other pesticides has existed for centuries. About 1000 BC, the Greek author Homer mentioned the use of sulfur for pest control. By 900 AD the Chinese were using arsenic to control garden pests. By the late 1600s, pesticides were used in Europe and later in the U.S.A. Both inorganic chemicals (sulfur, arsenic, mercuric chloride, carbon disulfide, etc.) and natural botanical substances were used (tobacco, pyrethrum, quassia, derris, etc.). The first really important synthetic organic hydrocarbon, DDT, was first made in the laboratory in 1874 in Europe. It was found to be a good insecticide and was used during World War II for human louse and mosquito control. With favorable publicity, its use increased greatly (over four times) after the war, in agriculture. Since then the genie of chemistry has truly been unleashed. At least a thousand new chemicals are synthesized and tested every year.

Mechanization. Modern farming would not be possible without the many and varied machines used for plowing, planting, cultivating and harvesting. Early milestones of mechanization include the invention of the cotton gin by Eli Whitney in 1793, the steel plow manufactured by John Lane in 1833, the horse-drawn grain reaper by Cyrus McCormick in 1834, the grain combine by J. Hascall in 1836, the steel share plow by John Deere in 1837, a practical threshing machine by the Pitts brothers in 1837, and the horse-drawn double-row corn planter by D.S. Rockwell in 1839. The first steam tractors were produced in the U.S. as early as 1849, and the first gasoline tractor was built in 1892 by John Froelich. The first gas tractor factory began in 1903, but tractors did not really completely displace animal power until around World War II (it wasn't until about 1950 that there were more tractors than horses on American farms).

Hybrids. Farmers and herdsmen have long improved their crops and animals by selecting the strongest or best-producing individuals for breeding future generations. In 1812, John Lorain in Pennsylvania demonstrated that crossing two varieties of corn resulted in greater yield. By 1850, many farmers all over the country were purposefully crossing different varieties. But the principles of genetics were not discovered scientifically until 1866, when the Austrian monk, Gregor Mendel, published his experiments with garden peas in an obscure journal. But Mendel's work went unnoticed by the scientific world until 1900. In the early 1900s the concepts of genes and mutations were announced by Thomas Hunt Morgan and Hugo DeVries, respectively. Also, in 1900, W.A. Orlon and E.L. Rivers developed a variety of cotton resistant to cotton wilt disease. From 1916 to 1918, Donald F. Jones developed modern corn hybrid methods (double crossing, using four inbred lines instead of two). In 1926, hybrid seed corn became commercially available, and it gradually displaced the use of open pollinated (non-hybrid) varieties, especially after World War II.

So we can see from these examples that modern agriculture resulted from a natural growth process that began many years earlier, and that it has "exploded" since World War II along with the rest of science and technology. The sharply increasing world population curve greatly added to the demand for food and fiber, and science stepped in to fulfill that need.

Spreading the gospel. The present state and federal agricultural institutions can be traced back to as early as 1776, when proposals for an agricultural branch of the government were first made. President George Washington recommended the creation of a national board of agriculture in 1796. Nothing was done until 1839, when Congress appropriated $1,000 of Patent Office fees for collecting agricultural statistics, conducting agricultural investigations, and distributing seeds, organized under an Agricultural Division of the Patent Office. In 1856, a 5-acre test garden was begun. On May 15, 1862, President Lincoln signed a bill creating an independent Department of Agriculture, headed by a Commissioner responsible to him. The Department was raised to cabinet status in 1889. The Morrill Land Grant Act of 1862 provided for the establishment of the state land grant colleges. The first state agricultural experiment station was established in 1875, the Connecticut Experiment Station (at Wesleyan University). The Hatch Act in 1887 provided a yearly federal grant to support an experiment station in each state.

In 1892, the cotton boll weevil invaded the U.S. from Mexico, and in ten years the U.S. cotton industry was in danger of going under. Research carried out by the USDA (United States Department of Agriculture) developed control measures involving planting so that the crop matured early and escaped most weevil attack, but few farmers adopted them. S.A. Knapp of the Bureau of Plant Industry conceived the idea of getting some farmers to try new methods on their own farms (under an indemnity fund to reimburse a volunteer for any losses suffered). This succeeded so well that in 1906, the General Education Board of New York City agreed to supplement government funds to employ field agents to work full time with farmers in one county. Thus was born our system of county agents, supported by government and state university research. In more recent years, considerable grant money has been given to universities to do research by the industries manufacturing agricultural products (fertilizers, pesticides, equipment, hybrid seeds, etc.). And of course, the whole agricultural journalism and advertising industry has grown up to disseminate information and promote products.

Marketing/distribution. At least as far back as the tin can (1818) and condensed milk (1856), there has been processed food, and the present marketing/processing/distribution system has developed since then. Other milestones along the way include the invention of the refrigerator in 1801 by Thomas Moore; efficient railroad transportation, which began to be developed in the 1840s; railroad refrigerator cars in 1851 and especially after 1868; the passage of pure-food laws by Illinois, Michigan, New Jersey and New York in 1881; the first USDA export food inspection (pork products) in 1890; approval of the federal Pure Food and Drug Act and Meat Inspection Act in 1906; the Cotton Futures Act, approved, 1914; the first attempt to regulate marketing of farm products, the fixing of a minimum price for a bushel of wheat ($2.20) by President Wilson in 1917; the approval of the Grain Futures Act in 1922; approval of the Agricultural Adjustment Act in 1933; and approval of acts incorporating the principle of flexible price support, change in the parity formula, and authorizing set-asides and subsidies in 1948–49 and 1954.

Worldwide system. Such are the beginnings of our present system. No farmer in this country is untouched by it. Most countries in the world have dealings with it, whether by importing our agricultural products or exporting theirs or buying machinery or growing the latest genetically engineered crops. The International Monetary Fund (IMF), World Bank, World Trade Organization (WTO), various American foundations and agencies and international corporations work together to promote and enforce the aims and methods of big agriculture.

Basic approach. But what are the fruits of modern agriculture? Are they all good—or are there undesirable side-effects? The basic philosophy or approach of this "scientific" system of agriculture seems to be to manipulate and control nature, to change natural systems to suit (and profit) man. Slick advertisements for herbicides and pesticides use dramatic, war-sounding slogans. Is this approach good?

There are established laws by which the natural world operates, which cannot be broken with impunity.

• The over-all ecological food chain operates with the scores of species of plants, animals, and microbes being in a delicate balance, the so-called "balance of nature." A disturbance of one level or species in the food chain affects other links in the chain.

• The various organs of a person's, animal's, or plant's body have definite functions, and under proper conditions they operate well, and the organism is healthy. Improper nutrition and environmental conditions can cause malfunctions, allowing disease microbes, parasites, and other pests to invade.

• Any individual cell of a person's, animal's, or plant's body is made up of dozens of parts (nucleus, genes and chromosomes, ribosomes, mitochondria, endoplasmic reticulum, lysosomes, etc.), plus thousands of enzymes, and energy-carrying "raw material" molecules. In a normal, healthy cell, all of this biochemical "machinery" works beautifully, performing the flurry of activities of any cell: breaking down food to produce energy, putting together components in growth and secretion activities, and in green plant cells, photosynthesis.

• The thousands of chemical molecules, ions, and atoms inside living organisms, in the non-living environment (rocks, soil, water, air), and in the whole universe, have definite properties (see any chemistry or physics textbook). This accounts for the distinct differences among, for instance, steel, paper, and water. And think of the different properties of ice, steam, and liquid water, all forms of H_2O.

Natural laws cannot be changed—or broken—without a penalty. The law of gravity, the speed of light, and the properties of oxygen simply can't be violated or changed. Science and technology in general, and the educational system they nurture, approach problems and develop new concepts with the philosophy of overpowering and changing nature to suit their purposes. If a mountain is in the way—move it! If an epidemic of grasshoppers is devouring crops—zap them with a poison! If corn won't grow on out-of-balance soil—develop a new variety that will!

Results. So with this in mind, let's look at some examples of the legacy of modern agriculture—the results of our "scientific" age of technology. Let's see if the results are desirable and for the good of mankind. To cite a couple of old sayings, "The proof is in the pudding" and "Nothing succeeds like success." How successful is modern science and its control-nature philosophy?

Vanishing soil. Soil is absolutely necessary for agriculture—fertile soil! Notwithstanding futurists' predictions (dreams) of hydroponic farming, the world is going to be fed with plants grown in soil for the foreseeable future. Some parts of the world were blessed with incredibly fertile soil, especially the United States of America. Crop yields in the days of the pioneers were every bit the equal of today's. According to the *Farmer's Every-day Book,* published in 1851, "Mr. J.P. Jones, of Sullivan Co., N. Y., in 1849, [raised] over 195 bushels of ears [of corn] per acre," while "Mr. Young, of Kentucky, in 1840, raised over 190 bushels per acre . . . in 1841, B. Bradley, Bloomfield, N. Y., raised 232 bushels on two acres" (p. 406–407).

But such yields did not always continue. Wrong farming practices often wore out land in a few years, so the family just packed up and moved west. Some farmers even spoke proudly of how many farms they had "farmed out" in their lifetime. But then by 1890, there was no more frontier—no more large areas of virgin land. Increasingly worsening soil loss in the late 1800s and early 1900s led to localized interest in soil conservation. World War I took over 50 million acres of European land out of production, so Americans patriotically plowed 40 million acres of the Great Plains while market prices rose. Wheat became a cash crop. New machinery produced more efficiency and higher yields, but after the war this led to a glut on the market and lower prices. So the "natural" thing to do was plow more land to grow more wheat to pay off debts. All this unprotected land, combined with devastating droughts in 1933 and 1934, spawned the dust bowl and darkened the sun as far east as Boston. This calamity plus the Great Depression produced a plethora of government New Deal agencies, including the Soil Erosion Service in 1935 and later that year the Soil Conservation Service. So began the government's predilection for paying farmers to do what the government wants—to install terraces, to grow soil-conserving crops, to not grow "soil depleting" crops, or to not grow anything at all! Today there are grants available to persuade farmers to switch to "conservation tillage."

Surely all those conservation programs and all that money have reversed the loss of our precious soil. No doubt they did for a while, but what do we have today? Spurred on by some of the same economic factors as earlier in the 20th century, farmers have plowed fencerow to fencerow—and then ripped out the fences—and windbreaks—to make room for gigantic tractors and center-pivot irrigation rigs as the topsy-turvy economy dictates "efficiency" and large acreage. Over-reliance on and over-application of humus-destroying and soil life-destroying synthetic fertilizers and pesticides have made soil sterile and like powder or concrete.

The results are astounding. The Soil Conservation Service estimated (1979) an average yearly erosion loss of 9 tons per acre, but on rolling Iowa cropland, about 40–50 tons per acre, and on unprotected hillsides, 200 tons per acre are going down our rivers! Under ordinary agricultural conditions, new topsoil can form at a rate of 1.5 tons (a layer 0.01 inch thick) per acre per year, but soil is being lost at a rate seventeen times faster than that. Obviously this soil loss cannot go on much longer. In some areas, the topsoil is already gone and farmers are farming subsoil. The water-borne soil clogs up our lakes, reservoirs, harbors, and ship channels, requiring millions of dollars of dredging costs. Fortunately, since 1980 cropland erosion has been cut considerably, from about 3 billion tons to less than 2 billion tons per year. Still, that is far too much. Dust bowl conditions have again started to develop in the Great Plains. In 1977, about 1 billion tons of soil were blown off of plowed fields. In 1979, 1.4 million acres were damaged by wind erosion in ten Great Plains states. Recent severe drought conditions west of the Mississippi River during the early 2000s have led to devastating wildfires and more soil loss. An estimate in 2000 said that up to 40% of global croplands have some erosion, loss of fertility or overgrazing.

Not only is our soil disappearing down our rivers and through the air, but nearly 7000 acres *per day* are being paved and gobbled up by creeping urbanization (about 2.5 million acres per year). An average traffic interchange takes 40 acres. Since 1945, an area larger than the state of Nebraska has been lost. About half of the time, cities and roads are built on good land, since there are fewer costs for leveling and grading.

There have also been shocking losses in soil structure and fertility. The USDA estimates that in areas where the topsoil is 10–15 inches deep, the loss of 6 inches of topsoil may reduce crop yields from 20 to 42 percent. Scientists warn of a worldwide threat from "desertification," the increase of deserts. In the last part of the 20th century, 8.6 billion acres of the world's crop and pasture land turned into arid deserts because of bad farming and grazing practices, improper irrigation, and forest destruction. This is an area about the size of North and South America combined! Africa and Asia are the hardest hit, but the problem is increasing in other areas, including the United States.

Adequate humus (from 3–10%) is necessary to give soil a loose, crumbly texture (allowing entrance of water and air, reducing erosion, and making soil easy to work), to serve as a storehouse of vital plant nutrients, and to buffer soil against toxic substances and extreme pH changes. Humus is produced by a host of beneficial soil organisms (bacteria, actinomycetes, fungi, worms). Besides producing humus, some of them assist plant roots in absorbing nutrients (mycorrhizae), others provide additional nutrients (nitrogen-fixing bacteria, mineral-releasing bacteria), some supply plants with growth-stimulating substances (hormones and vitamins from bacteria and fungi), and some actually protect plants from diseases and pests (by producing antibiotics and by suppressing or destroying harmful microbes, worms and insects).

Yet some commonly used fertilizers (anhydrous ammonia and high-salt fertilizers), as well as most pesticides and herbicides, kill, inhibit or disturb the natural balance of this volunteer work force in the soil. High ammonia fertilizers actually make humus soluble so it can readily leach away. Humus cannot be produced if the "humus factory" is destroyed (soil life). Also, the tendency away from most farms having livestock toward all-grain farming (and over-concentration of livestock and poultry in large feeder operations) means there is little or no manure being recycled back into the soil to provide adequate humus (recycled crop residues and green manures do help, but animal manures used in addition are much better). Continual removal of crop matter (grain, stalks, hay) and insufficient replacement means that humus levels will soon drop too low to provide adequate soil fertility. Science and the von Liebig mentality say that the lack of fertility can be made up totally by inorganic synthetic fertilizers. But many biochemical studies have shown that this simply is not true. Yields in bulk (quantity) may stay high, but *quality* and nutritional value of crops suffer immediately, and even bulk yields will eventually decrease, at least after it becomes economically prohibitive to keep pouring on the chemicals at increasing rates.

Floods. Did you realize that topsoil and humus deterioration have been a major contributing factor in the devastating flooding that has occurred in recent years? In their book, *The Flood Control Controversy,* L.B. Leopold and T. Maddock, Jr. say: "The incorporation of humic materials into the upper horizons [of the soil] and the consequent association of soil organisms promote the development of soil structure and good tilth which make the soil more permeable. Furthermore, organic material at the soil surface physically aids in diffusing water over the ground in numberless microchannels, and then, by preventing high velocity of flow, inhibits the cutting of rills or channels. This action on velocity increases the infiltration opportunity and also contributes to detention storage during runoff" (p. 59).

Another quote: "Deterioration of tilth and structure lead to reduced water infiltration and water-holding capacity, reduced natural fertility, and increased erosion rates. If measures are not taken to restore the soil to its proper structure, tilth, and organic matter, their loss will have a snowball effect. Soil fertility and the infiltration rate will decrease, and organic matter content will decrease even more" (1981 USDA Yearbook of Agriculture, p. 144).

According to USDA figures, soil with a 4 to 5% organic matter content can absorb a 4–6 inch rainfall in an hour, while soil with a 1½–2% organic matter content can only absorb a ½–1½ inch rainfall. Typical U.S. corn belt soils are down to only 0.5–2.5% organic matter.

Just about any soils textbook will have a graph showing what happens to soil organic matter under typical agriculture. Organic matter is lost, at first rapidly, then more slowly. Tests at the Missouri Agricultural Experiment Station found organic matter losses of about 25% the first 20 years of cultivation, 10% the next 20 years, and 7% the third 20 years.

S.G. Archer says it eloquently in his book, *Soil Conservation*:

> The nation's river systems are rich, and some are dying. The trouble stems from man's use of the land. Virgin forests and verdant native grasslands that once protected mountains and plains are gone or are badly depleted. The protective forest canopy, with the leaf mold beneath, and the tall luxuriant grass protected the soil from the devastating power of the falling raindrops and held runoff water back until most of it soaked into the ground, protecting the soil while the surplus water moved slowly off to clear streams. That was the situation before the white man had dominion over the soils of this continent.
>
> Today, fifty to three hundred years after man began to conquer the land with axes, plows, sheep, and cattle, there is little natural protection for the soil, and there is even less planned protection established by the most successful group of woodsmen, farmers, and ranchers the world has ever known. The Midas touch of the conquerors of the land turned natural resources of grass and trees into money.
>
> But the land is bare.
>
> Many of the nation's forests have been ravished, and most of the rest have been impoverished. The forest land produced crops for a while, and then gullies, where torrents of muddy water, unhindered, carried the remaining soil to creeks and rivers. The grasslands were either plowed or overgrazed, often with the same erosive effects. But even where erosion did not develop, the grass lost its ability to hold back the rainfall and to let it soak into the ground.

Cropland too often is tilled, planted, and harvested with never a thought of protection or of the damage being done to the structure of the soil. The rain packs the unprotected soil instead of enriching it, thus permitting less and less of the rain to go into the soil and more and more to run off, carrying with it huge quantities of topsoil to be deposited on the bottom lands and in the channels of creeks and rivers.

As man's dominion of the land became more complete, floods and flood damage increased in frequency and intensity.

Great avalanches of mud move into and down all of the nation's river systems, carried by more and more water as the soils, the ranges, and the forests deteriorate.

Rivers and creeks which once carried a steady flow of water throughout the year become raging torrents during heavy rains and then dwindle to trickles during the dry months, because the rainfall runs off instead of soaking in to renew the underground watercourses which used to feed the streams during the dry season.

The loss of lives in floods is recorded as news. Damage to property, especially in cities, makes the headlines when a great river goes on a rampage. Losses of crops are sometimes recounted, but the damage to the land by scour and silt deposition is seldom mentioned. In newspapers, floods are generally attributed to big rains upstream, but the contributory factors of range, forest, and cropland deterioration are not recognized (p.199–202).

Yields. USDA studies have shown that loss of topsoil reduces crop yields. The loss of an inch of topsoil can reduce corn yields by 3 to 4 bushels per acre, and if the topsoil is only 10–15 inches deep, the loss of 6 inches can reduce yields by 20–42%.

Beginning in the early 1970s, yields of most U.S. crops hit a peak, and then began falling, dipping up and down in succeeding years, or else leveled off. Corn yields peaked in 1972 at 97.1 bushels/acre then fell to 71.3 in 1974 (because of widespread drought) before inching back up to 91.5 in 1977. There have been some increases since then, around 115–120 bushels/acre in the 1990s and over 150 in 2004. Wheat yields peaked at 33.9 bushels/acre in 1971, were pretty stationary at less than 31 for awhile, and since then have risen to 40–45 bushels/acre in good years. The experts do not know what is causing yields to stop growing at the rates they had been since the 1940s. Sylvan Wittwer of Michigan State University speculates that yields may be faltering because soils are losing organic matter, or are being compacted by heavy machinery, or because air pollution is harming crop growth, or because fewer pesticides and high-priced fertilizers are being used, or because less-productive land is being cultivated.

Water too. An acre of corn requires about 500,000 gallons of water during a growing season. Agriculture accounts for 96% of the water consumed in the U.S. About 12% of U.S. cropland is irrigated. In the water-short western states, irrigation uses about 80% of all water used. We are "mining" water at such a great rate that the mammoth Ogallala aquifer, an underground sea of water stretching from South Dakota to the Texas panhandle, is dropping about 10 feet a year, while the groundwater is being replenished at only about 3 inches a year.

Another problem associated with irrigation in dry climates is the build-up of salts at or near the surface because of high water evaporation rates from the wet soil. Also, the salinity of some rivers used for irrigation is increasing, such as the Colorado River, whose water is 18 times saltier near its mouth than at its headwaters. Many acres of once-lush farmland have been turned into a salt desert.

Chemicals to the rescue! Chemicals to control crop and hay insects costing $306 million were used on 42 million acres, and herbicides costing $356 million were used on 89.9 million acres in 1969, while in 1978, treated acreage increased to 75.7 million acres for crop and hay insects and 163.5 million acres for weeds. This doesn't include millions of dollars spent for chemicals to treat livestock and poultry, nematodes, crop diseases, and for defoliation ($121 million in 1969 and $166 million in 1974). In 2000, nearly 540 million pounds (active ingredients) of pesticides were used in the U.S., of which 57% were herbicides, 14% insecticides, 7% fungicides, with 22% as minor categories. In 1936, only about 30 pesticides were registered in the U.S., while by 1971, over 900 were registered.

Pesticide usage is varied. About half of insecticides are used on the nonfood crops of cotton and tobacco, while the food crops corn, fruits, and vegetables receive a third. Nearly half of all herbicides are used for weeds in corn, by far the largest amount for any crop. Fruit crops consume the greatest amount of fungicides, 60 percent. On an acreage basis, omitting pastureland, 34 percent of U.S. cropland is treated with herbicides and 12 percent with insecticides.

Results? What have been the results of this outpouring of mainly synthetic chemicals on the land? Currently an estimated 33% of all crops is lost annually to pests (13% to insects, 12% to pathogens, and 8% to weeds) in spite of the use of pesticides and nonchemical controls. Crop losses due to insect pests have increased nearly twice from the 1940s to 1974 (from 7% to about 13%), in spite of a ten-fold increase in insecticide usage. In the same period, losses from plant diseases and nematodes increased slightly (from 10.5% to about 12%), while losses from weeds declined (from 13.8% to 8%). Loss figures in the U.S. for 2003 include nearly $3 billion per year, just from the Colorado potato beetle, corn rootworm, diamondback moth, silverleaf whitefly and green stinkbug.

"The substantial increase in crop losses due to insect damage despite increased insecticide use can be accounted for by some of the major changes that have taken place in agriculture since the 1940s. These include the planting of some crop varieties that are more susceptible to insect pests; destruction of natural enemies of certain pests, which created the need for additional pesticide treatments; increase in pesticide resistance in insects; reduced crop rotations and crop diversity—with an increase in continuous culture of a single crop; reduced FDA tolerance and increased 'cosmetic standards' of processors and retailers for fruits and vegetables; reduced sanitation, including less attention to the destruction of infected fruit and crop residues; reduced tillage, with more crop remains left on the land surface; culturing crops in climatic regions in which they are more susceptible to insect attack; and use of pesticides that have been found to alter the physiology of crop plants, making them more susceptible to insect attack" (from an article by David Pimentel and other authors in the journal *Bioscience,* 1978).

Read over that paragraph again. It's important. What Dr. Pimentel and his associates are saying is an indictment against many of the methods of modern agriculture, most of which are designed to produce greater yields more efficiently ("bigger is better") and to control nature.

What else? And what other side effects have there been as a result of increased use of agricultural chemicals? Few people are unaware of the growing menace of environmental pollution and the upset ecosystem. A few examples will suffice:

1. In 2000, an estimated 3 million individuals were poisoned by pesticides globally.

2. Because of their widespread use, pesticides are consumed by most people. An Environmental Protection Agency study found from 93 to

Ladies and/or Gentlemen

We know that the bodies of humans and many animals, even insects, have hormone-secreting organs, altogether called the endocrine system, whose special chemicals, called hormones, regulate many different body functions, among them metabolism, growth and reproduction. Most plants' growth and reproduction are also regulated by hormones (which are chemically different from animal hormones). You also may know that if a child is deficient in a sex hormone, their maturation will be slow or abnormal.

Disturbing clues. Beginning in the late 1940s and 1950s, scientists noticed that certain animals began having trouble reproducing, with population levels dropping: bald eagles in Florida, otters in England, sea gulls in southern California, and famously, alligators in one lake in Florida, where the males had abnormally small reproductive organs, and eggs didn't hatch or the young died within two weeks (just before that, there was a pesticide spill in the lake—dicofol, a DDT relative). In the late 20th century, reports began to accumulate of early puberty in girls and boys from around the world, in 7, 8 or 9 year-olds, and sometimes even in 2 or 3 year-olds. A Danish reproductive biologist has found that between 1938 and 1990, the average human sperm count fell nearly by half. What is going on?

Tracing the cause. The early cases of reproductive problems in wildlife pointed toward newly-introduced pesticides—DDT and dieldrin. Later other synthetic chemicals were suspected, polychlorinated biphenyls (PCBs), and the synthetic hormone diethylstilbestrol (DES), commonly used to promote growth of mass-produced chickens and cattle, and given to humans for difficult pregnancies, birth control and prostate cancer. It is well known that DES is a synthetic form of the female hormone estrogen.

The way our natural hormones work is by the characteristic shape of their molecule fitting into special hormone receptor sites (like a key in a lock), either on the outside or inside of certain body cells, triggering the cells' DNA (genes) to produce proteins that affect the reproductive functions (or whatever function that the particular hormone controls).

If the synthetic DES hormone can mimic real estrogen, can other chemicals which aren't intended to, do the same? Much research and many sad cases have answered, yes! And usually in extremely small concentrations, such as we can unknowingly ingest in food and drink, or even absorb through our skin. They are called *endocrine disrupters* or *hormone mimics,* and their potential harm to most life on earth is tremendous. Some don't increase hormone functions, but block them instead.

Since the chemical industry began tinkering with molecules, over a century ago, well over 100,000 synthetic chemicals have been manufactured, and about a thousand new ones are introduced every year. Unfortunately, most of them are unleashed upon us without adequate screening and testing. Since plants, animals and humans have had no experience with them during their history, it is not surprising that some synthetic chemicals will have very negative effects on life.

The culprits. Following are *a few* of the nearly 1000 dangerous hormone-mimicking chemicals that have been pin-pointed so far. Not all of them have direct agricultural connections, but we all should be concerned.

1. *Pesticides.* Atrazine (herbicide, causes abnormal gonads in frogs and alligators); dieldrin, methoxychlor, pentachlorophenol, methyl parathion, DDT (insecticides, cause sterility and mutations in male rats, block nitrogen-fixing bacteria on legume roots); and vinclozolin (fungicide, causes low sperm production in rats).

2. *Plastics.* Plastics of many kinds are all around us. Some of them are made using chemicals called plasticizers, including some called phthalates, bisphenol-A, DEHA, and another, p-nonylphenol (found in polyvinyl chloride [PVC], polycarbonate and polystyrene plastics). They can cause growth of breast and uterine cancer cells. The hormone mimics can leach out of the plastics under certain conditions. Dangerous plastics are found in food and drink containers, plastic wrap, linings of metal food cans, carpet backing, shoe soles, garden hoses, toys, fake leather, and a multitude of other household and everyday products.

3. *Industrial chemicals.* Some well-known toxic chemicals also act as endocrine disrupters, including over 200 varieties of PCBs, 75 types of dioxin, 135 kinds of furan, and brominated biphenyls. Some of these chemicals have caused abnormal embryo development in fish.

4. *Agricultural hormones.* Hormones are (or used to be) administered to livestock and poultry, mainly to stimulate growth (including DES, see above) or to increase milk production (rBGH, genetically-engineered bovine growth hormone). Excess amounts of these hormones can be found in meat and milk, and they may be one cause of early puberty in children. Now that this problem is well-known, more and more farmers are eliminating hormones.

5. *Other common chemicals.* Fabric stain protectors usually contain fluorinated surfactants which can volatilize and become perfluorocarboxylates and other endocrine disrupters. They have been found in arctic seals and polar bears. Alkylphenol polyethoxylates are found in some detergents, pesticide sprays and personal care products. They have been used since the 1940s and can break down into other estrogen-mimicking chemicals.

100% of the people tested had one or more pesticides in their bodies. About 50% of foods sampled had detectable residues of insecticides. "Unfortunately, little is known about the effects long-term, low-level dosages of pesticides may have on public health. Furthermore, the possible interaction between low-level dosages of pesticides and the numerous drugs and food additives the public consumes has not been fully studied," according to David Pimentel in *Pest Control Strategies*, 1978.

3. Chemical control for U.S. crop production was estimated in 1978 to be $2.2 billion annually. But this did not include what are called "external costs" of at least $8 billion annually. These include (a) hospitalization costs for 6,000 human pesticide poisonings, (b) about 60,000 days of work lost from these pesticide poisonings (money costs of about 200 deaths were not included), (c) medical costs for an additional 8,000 human pesticide poisonings treated as outpatients, (d) about 30,000 days of work lost from people not ill enough to be hospitalized, (e) about $12 million in direct honey bee losses, (f) reduced fruit crops from reduced pollination

because of destruction of honey bees and wild bees, (g) livestock losses, (h) commercial and sports fish losses, (i) bird and mammal losses, (j) destruction of natural enemies of pests, resulting in outbreaks of other pests, (k) pest problems resulting from pesticide effects on the physiology of crop plants, and (l) increased pesticide resistance in pest populations.

4. Over-use of commercial fertilizers and poor manure management, plus high levels of soil erosion mean that considerable amounts of fertilizer nutrients wind up in our rivers and lakes. These nutrients stimulate the growth of water plants (algae, "water weeds"), which then undergo a "population explosion" (called a bloom), sometimes clogging commercial and sporting waterways. And when they die, their decomposition uses up the water's oxygen and produces toxins which kill fish and other aquatic life. This process of increased growth and fertility of aquatic habitats is called eutrophication. It has become a serious problem since the late 1940s. For a while, it looked as if Lake Erie would "die." During the period 1948–1962, nitrates in Lake Erie increased 50%, while phosphate increased over 400%.

The soil under long-term livestock feedlots has been found to contain 2000–4000 pounds of nitrate nitrogen per acre. High nitrates in groundwater and drinking water is a common phenomenon in agricultural states, and now pesticides are a common pollutant of groundwater. In 1980, a Wisconsin study found 311 of the state's 11,396 small noncommunity water systems had more than the allowable concentration of nitrates, and a 1982 study found synthetic chemicals in 51 of the 208 community wells tested.

Groundwater is our most fragile resource. Since groundwater moves slowly in tiny soil and rock crevices, contaminated groundwater is practically impossible to clean up. Stopping the source of pollution and allowing natural water to be added to the groundwater to dilute the polluted water would take decades. Polluted water pumped to the surface can be purified, but at great added expense.

The verdict. The evidence is clear—and is mounting daily—that unwise and excessive use of agricultural chemicals (plus the additional sources of industrial and urban chemicals and wastes) is horribly polluting and poisoning our environment—and us. It can't go on much longer!

Weeds too. Not only has spraying millions of pounds of insecticides failed to keep the bugs under control, but problems are increasing with herbicides and weeds. In a speech to the North Central Weed Control Conference in December 1980, G.F. Warren of Purdue University admitted,

"We have seen the rapid increase in resistant weed species almost from the beginning of commercial application of selective herbicides." Weed species which never were a problem in our grandfather's day have now invaded our prime agricultural land. Herbicides that used to kill weeds now are either ineffective or have to be used at increased rates. Pesticide expert George Ware of the University of Arizona says that our pesticide arsenal is not keeping up with our pest problems because of resistance. "This health-conscious generation knows the undesirable effects that chemicals in excess—not just pesticides, all chemicals—can have on our environment. And more and more our farmers are beginning to know what their cost-benefit ratio is . . ." (*Pesticides: Theory and Application,* 1983).

Less variety. Another phenomenon of our modern agricultural system is the decreasing variety and number of crop plants. According to J.R. Harlan, "most of the food for mankind comes from a small number of crops and the total number is decreasing steadily." Botanist P.C. Mangelsdorf lists 15 crops that feed the world: wheat, rice, corn (maize), sorghum, barley, beans, peanut, soybean, sugarcane, sugarbeet, sweet potato, manioc, potato, banana, and coconut.

According to 1969 figures, there are 269 varieties of wheat, but only 9 varieties comprised 50% of the U.S. wheat acreage. Only 6 major varieties of corn (maize) accounted for 71% of the planted acres, even though there are at least 197 commercial inbred varieties. Only 4 out of 82 varieties of potato were planted on 72% of potato land. About 97% of U.S. cotton acres are planted with one type: short-staple upland cotton, even though there are thousands of locally-adapted wild and hybrid varieties worldwide. And so on for other major crops.

In the 20th century, over 7,000 varieties of apples and 2,500 varieties of pears were grown in the U.S. Today, over 85% of those apples have disappeared, and only two varieties make up over half of the country's apple crop. Beginning in the late 1960s and early 1970s, many native varieties of crops were replaced by the high yielding hybrid varieties of the "Green Revolution," as well as hybrids in general. In 2003, 40% of U.S. corn acreage was planted with genetically modified seed, while 75% of soybean acreage and 73% of cotton acreage were planted with biotech seed, from a small number of sources. Genetically engineered crops were planted on 63% of U.S. acres used for crops. Also, increased development, urbanization, and destruction of natural habitats all around the world have threatened many wild varieties of crop plants. The U.N. Food and Agriculture Organization (FAO) says that 75% of the genetic diversity of crops worldwide was lost during the 20th century. Other estimates range as high as 95%. The many varieties of the crop species are a reservoir of genetic traits, some of which could potentially provide valuable characteristics for future crop improvement, such as increased food value, or disease resistance, or drought tolerance.

If those crop varieties become extinct, the potentially useful genetic traits they carry will be lost. Many crop scientists and botanists are worried about this "narrowed genetic base" and are establishing seed "banks" to preserve as many varieties as possible. Planting large acreage in only one or a few varieties can cause other problems, such as allowing rapid spread of pests or diseases to which the variety is susceptible. This is best illustrated by the disastrous wildfire spread of southern corn leaf blight in 1970, when 80% of the U.S. corn crop was planted in hybrid varieties susceptible to the disease.

Hybrids and biotech. Hybrid varieties are probably man's first "high tech" achievement in agriculture, although the earliest hybrids were made before the principles of genetics were understood. A hybrid is the offspring that results from crossing two or more different strains or varieties. The hybrid contains a combination of genetic traits or characteristics from the parents. By applying the laws of genetics, breeders can predetermine the characteristics of the hybrid, such as yield, height of plant, and resistance to a certain disease. In the more recent techniques of genetic engineering or gene modification, individual genes can be inserted as desired, even from other species of plants or animals. These man-designed varieties are uniform for certain characteristics, which is supposed to be an advantage; thus all the plants grow to the same height, mature at the same time, and produce uniform grain.

But the trouble is just that—they are all alike for those traits—there is no variation such as there is in open pollinated (non-hybrid) varieties. The individual plants of open pollinated varieties have slightly differing characteristics, which gives the total population of plants more versatility in withstanding environmental stresses and producing a good yield under nearly any conditions. On the other hand, hybrids are so "fine-tuned" to produce well under a certain limited set of conditions (soil type, fertility, weather), that if conditions are unfavorable, the crop may be a disaster. Also, hybrid seeds do not "breed true"; that is, the desirable traits of the hybrid appear uniformly only in the first generation. If the seed is saved and planted the next year, only a small percent of the plants will be the original hybrid, and most will have undesirable traits. Thus, the farmer must buy new hybrid seed every year—he is at the mercy of the seed companies.

Furthermore, the particular genetic traits which the scientists "build in" to the hybrid are not always wisely chosen. Generally, the number one emphasis is YIELD; that is, bulk—bushels, tons. Much less attention is paid to QUALITY and nutritional value. Breaking yield records is much more important than growing high quality food that animals or people can thrive on. Amino acid, mineral, and vitamin content are way down on the list.

Tests of hybrid corn varieties by Ernest M. Halblieb of McNabb, Illinois, revealed that the hybrids did not take up seven to nine trace minerals. None absorbed cobalt, a mineral needed by animals in vitamin B_{12}. Protein content of wheat and corn steadily declined during the 20th century.

Going green? During the late 1960s the so-called Green Revolution burst upon the scene, at first with "miracle rice," hybrid varieties touted to be high-yielding and the solution to hunger and poverty in developing countries. Many third-world governments promoted them. The only catch was, the genetic traits programmed into the rice, and later wheat and corn, did yield from 10 to 100% higher than local varieties . . . but only if the plants received high inputs of fertilizer, herbicides, pesticides and water, most of which cost poor farmers so much that they either could not afford them or else had to rely on government subsidies.

Nevertheless, Green Revolution hybrids took over hundreds of millions of acres in struggling countries through the 1970s to 1990s, causing small farmers to abandon their traditional seed varieties and methods, making huge areas more vulnerable to crop pests and diseases due to lack of genetic diversity, and making farmers and governments dependent on big agribusiness corporations. In most developing countries, there was some increase in food production, but an approximately equal increase in hunger. With all its drawbacks, the Green Revolution has turned brown.

The next big thing—GMOs. Producing new hybrid crop varieties by traditional methods is an expensive, time-consuming process (often ten or twenty years), because thousands of genes are involved in crossing plants using pollen and eggs, with precise combinations of traits from two parents requiring many generations of previous inbreeding to obtain pure lines. So in the 1970s and '80s, scientists developed a much faster, more efficient way of modifying both plants' and animals' genetic makeup, called genetic engineering (GE) or gene modification (GM).

Using various high-tech methods, including bacteria or viruses as carriers and "gene guns" to blast them in, specific genes from virtually any organism can be inserted into cells of just about any other organism. Just to see if it could be done, scientists have put genes from a fish that lives in freezing Antarctic waters into tomatoes to make a frost-resistant crop. They have used genes for bioluminescence from jellyfish to make pigs and tobacco that glow in the dark! They are trying to re-engineer photosynthesis in rice in the hope of increasing yields by 50%. By inserting certain microbial genes into crops, cows or pigs, these organisms can be made to produce antibiotics, hormones, vaccines or other pharmaceutical drugs, a process often called "pharming."

Farmers are probably aware that there are commercially available genetically modified (GM) varieties of corn, soybeans, rice and cotton, along with dozens of others being used or in development. Mainly they are either herbicide-resistant or resistant to a few insect pests (from the *Bt* gene, found naturally in a common soil bacterium). GM crops are not usually bred for higher yields, so any yield increases they may produce are due to their resistance to pests and herbicides (better weed control); in fact, after herbicide resistant soybeans were introduced in the U.S., average yields were 5–10% lower than with non-GM soybeans.

The adoption of GM varieties by U.S. farmers has been surprisingly high; in 2004, 85% of soybean acreage, 45% of corn land and 76% of cotton fields were planted with GM seed. Worldwide around 170 million acres have GM crops, with the U.S. planting 63% of GM acres, Argentina 21%, Canada 6%, Brazil 4%, China 4% and South Africa 1%. We are definitely in the middle of a GMO (genetically modified organism) wave. What will the future bring?

The system. But that's part of the system—the establishment—the interconnected association of government, industry, and education that runs our society, agriculture included. Much evidence shows that in recent decades, the system is not being run to benefit the farmer. Government programs and policies are oriented toward providing the urban consumer with low-priced food ("cheap food policy") and interconnecting the world's exporting countries by trade commitments ("one worldism," free trade, globalism).

Market prices for farm commodities are manipulated to force farmers to increase production and become "more efficient" (larger farms, more mechanization, higher yield per unit cost). Interest rates and costs of machinery, fuel, and fertilizers are now extremely high, which forces the smaller, "inefficient" farmer out of business. U.S. farm and economic policies are set by politicians, economists, bureaucrats, businessmen, and scholars—NOT FARMERS! The President's Council of Economic Advisers, the Committee for Economic Development, the Council on Wage and Price Stability, the Joint Economic Committee of Congress, and the Federal Reserve System are in charge—NOT FARMERS!

In a 1962 document titled "An Adaptive Program for Agriculture," the Committee for Economic Development, a prestigious group of 200 leading corporate businessmen and scholars, spelled out a long-range program to move people and land out of agriculture by discouraging young people from entering farming careers, providing urban jobs for farmers, and manipulating (lowering) the price support system to discourage farmers from farming. Their program is succeeding very well.

So there you have some of the fruits of our "modern" agricultural system. Not very appetizing, are they? In fact, they are rotten! *Our modern economic and agricultural system is on the verge of collapse—if not in the process of collapse!* There HAS to be a better way. There IS a better way, and it's becoming more and more common.

Chapter 2

View from the Mountain —

ECOLOGY

B efore assaulting problems, it is best to back off a distance and be sure we comprehend the total situation—that we see the forest and not just the trees. Then we can better recognize problems, symptoms, causes, and solutions. So the first stop on our journey will be to climb an imaginary mountain and take a panoramic view of nature.

Agriculture works with natural systems, with living biological species of plants and animals, and with the natural environment. The branch of biology which deals with all of these things at once is *ecology*, the study of organisms (living plants and animals) and their environment, AND the interrelationships among them.

Nature is incredibly complex. It is composed of millions of functioning parts. Even one cell from any plant or animal is more complex than any man-made machine. Ecology studies organisms and their environment— how they live and function in their environment and how they interrelate with each other and with the environment. We find that these "parts" of nature, like the gears and levers of a machine, have their own functions and that all the parts together form a larger entity, a natural or biological *system*, or to use the terminology of ecology, an *ecosystem*. An ecosystem and its parts operate under certain natural laws. These established laws operate without fail, and if they are broken, a penalty results, just as when a criminal is punished for breaking a man-made law. If we understand more about natural systems and how they operate, we can often eliminate, or better yet, avoid problems (which are the penalty of broken laws).

Ecological Services

One concept of ecology that has become prominent in recent decades is the fact that natural ecosystems (plants, animals and microbes) provide a variety of life-sustaining functions or services for human societies. These ecological services include: 1. soaking up precipitation and releasing it into streams slowly, thus preventing flash floods and drying up of streams; 2. cleaning surface and ground water as it percolates through soil and marshes; 3. moderating weather extremes, as vegetation releases humidity into the air and prevents temperature extremes; 4. reducing plant and animal disease outbreaks, through disease-fighting mechanisms and increasing environmental complexity, which slows disease spread; 5. removing carbon dioxide from the air and trapping it in plants and healthy soil; 6. providing natural fertility for crops and all plants through the complex interrelationships of the soil, air, rocks, roots and microbes; and 7. pollinating many crops.

The increasing impacts of human activities, including conventional agriculture, have drastically reduced the ability of natural ecosystems to function by one-third to one-half. Such activities as clearing forests and filling in wetlands, using toxic pesticides and dumping toxic wastes, replacing natural habitats with farm fields and urban development, spewing carbon dioxide and toxic emissions into the air, and overfishing the oceans are seriously crippling the earth's capacity to support us. With the present 6.4 billion human population expected to grow to 9 or 10 billion by 2050, the pressures may become too much. Governments are focusing on supplying short-term needs and are ignoring long-term consequences. What can be done?

Far-sighted experts recommend increasing agricultural production per acre by optimizing safe use of fertilizers and other technology so fewer acres will have to be farmed, increasing soil organic matter so as to increase water and nutrient efficiency, increasing urban green space, maintaining enough natural or wild habitats to prevent loss of species (biodiversity), and using sustainable harvesting methods for forestry and fisheries. They especially point out the great need for enlightened policy making, with a global and long-term outlook.

As Gary Zimmer said so well in *The Biological Farmer* (2000, p. 27): "We could say that whenever a part of the natural soil-plant system is eliminated (such as earthworms killed by pesticides or plant matter harvested), the farmer has to take over the job of the part eliminated if he still wants to grow high-yielding, healthy crops. If nutrients contained in crops are removed and not recycled into the soil, the farmer has to replace them with expensive purchased fertilizers. If earthworms are not there to mix and aerate the soil, the farmer has to till it. If natural pest and disease control agents are crippled, the farmer has to use expensive and dangerously toxic chemical controls."

The food chain. The most important natural system to give us an overview of agricultural problems is the general food chain, the cycle of energy transfer involving stored energy called food. The basic cycle is very simple, and virtually everyone is at least subconsciously aware of it. Let's start with the organisms we are most familiar with—ourselves, and our domesticated animals. What do we use for food? Most people eat a variety of plant and animal food. What do our domestic animals eat? Plants, almost entirely. What do plants live off of? They make their own food, by the process called photosynthesis, from carbon dioxide, water, small amounts of mineral nutrients from the soil, and light energy trapped from the sun. Where do plants grow? In the soil, from which they obtain water and minerals.

These statements probably aren't much of a surprise. You probably learned them in grade school. We have now built up several steps of our food chain:

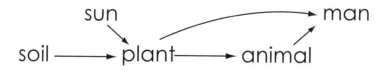

These are the most obvious steps and the ones most people, including farmers, are more or less familiar with. But are we finished? Could these parts operate continuously like that? No, we must add several more steps in order to complete the picture, like a belt to connect two wheels in a machine. Animals produce what is usually called "waste," mostly manure, plus urine. Most crop plants produce unused stalks, leaves, and roots, which we often call crop residues. These plant and animal "wastes" are not, or should not be called wastes, for in nature they are recycled. These materials, manure and plant residues, are called *organic matter*. They should be returned to the soil, where they should decompose, or rot, into a substance called *humus*. This decomposed organic matter is a natural plant food, supplying needed nutrients in the forms and at the time plants need them.

The final link in the chain is a multitude of mostly microscopic organisms in the soil which perform the functions of decomposing ("eating") fresh organic matter to form humus and helping to supply nutrients to plants. These soil microorganisms are a vital but generally overlooked part of the ecosystem. Thus we have built our food chain:

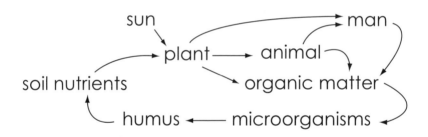

It is a system, a *cycle*. It "runs" on energy just as a man-made machine does, originally energy trapped from the sun by plants, and then passed on to each step in the system. It operates under definite laws. A deficiency in or elimination of one step in the system will decrease or eliminate the next step—and the next—and the next. In other words, the "balance of nature" is upset. And how easily man can throw the system out of balance! In fact, scientists around the world now warn that human alteration and degradation of natural ecosystems, such as wetlands, forests, estuaries and even worldwide ocean fisheries, have crippled about two-thirds of the earth's ecological "machinery" that ultimately supports life on earth. Such functions as cleaning freshwater, cycling nitrogen and decomposing wastes are being impacted.

Ecology out of Whack

Human populations and technological abilities have increased so greatly since the 18th century that it is becoming easier and easier for man's activities to drastically affect the earth's ecosystems. When the human population was small and technology simple (bows and arrows, compared to nuclear bombs), nature could easily absorb insults from human activities. A tree cut down here, a stream dammed there . . . no great damage. Storms and other natural disasters caused far greater destruction, and the earth's biological and physical systems repaired the damage.

But then after the Industrial Revolution in the late 18th century, people began mining more coal, polluting the air and dumping toxic wastes. In the latter half of the 20th century especially, man's assault on nature has been devastating. Let's list several examples of human environmental blundering.

1. *Poor agricultural methods.* For thousands of years, such practices as plowing hillsides, leading to soil erosion; overgrazing by livestock, causing elimination of grassland and overgrowth of scrub and desert vegetation; and irrigation in regions where salt build-up can occur, have damaged natural ecosystems and led to reduced agricultural production.

2. *Deforestation.* Elimination of forests over large areas can lead to soil erosion and flash floods, mudslides, loss of native plant and animal species, local change of climate (since forests release a lot of humidity into the air), and seasonal drought (since forests release rainfall slowly into streams). Cutting of forests by ancient cultures has led to the mostly denuded landscapes of Greece, Israel and Scotland. Recent clear-cutting of tropical forests for lumber and clearing of agricultural land has caused incalculable ecological damage, plus the poor tropical soils soon become unsuitable for crops. Recently, most species of eastern North American songbirds have declined in numbers alarmingly, partly because of deforestation, both in the U.S. and in the tropical habitats where they spend the winter.

3. *Alien invaders.* When people migrate to new regions of the earth, they usually bring plants and animals from their original

homeland, either on purpose or accidentally. If the foreign species survive, they often multiply unchecked because the predators and diseases that kept their populations in balance are not present. Native species are often exterminated and the alien species may become noxious pests. Of course most U.S. agricultural crops are from other regions (such as wheat, oats, barley, soybeans, apples, peaches, etc.), but there are hundreds of notoriously obnoxious imports, including quackgrass, Russian thistle, bindweed, leafy spurge, velvetleaf, Argentine fire ants, gypsy moths, Japanese beetles, European corn borer, Dutch elm disease, zebra mussels and jumping catfish.

4. *Dead zones.* Overuse of crop fertilizers in the U.S. corn belt and Lower Mississippi Valley has resulted in excess nutrients washing down the Mississippi River into the Gulf of Mexico. Since 1960, every summer populations of one-celled ocean algae multiply explosively until their death and decomposition uses up the water's supply of oxygen, killing virtually all fish and marine life. In recent years this "dead zone" off the Louisiana coast has been the size of New Jersey, and other dead zones have appeared in other regions.

5. *Climate change.* The earth's climate has constantly changed for millions of years, sometimes drastically (think of the tropical conditions of the "coal age" compared to the "ice ages"). These ancient climate changes took usually thousands of years to occur. But in the last 300 years (the Industrial Age) and especially since about 1910, the average world temperature has shot up like a rocket, although from 1940 to 1970 there was a temporary cooling. Droughts and heat waves are becoming worse and more common. Nearly every year seems to be a record-setter for heat. Glaciers and north polar ice are melting fast. Although some people refuse to acknowledge it, nearly all scientists who have studied it believe that global warming is real.

And the major cause of today's climate change is human activity, since temperature rise is closely correlated with so-called greenhouse gases, especially carbon dioxide (CO_2), but also methane (CH_4), nitrous oxide (N_2O), low-level ozone (O_3), and man-made chlorofluorocarbons (CFCs). These gases trap more of the sun's heat (infrared radiation) than can escape back into space, like a greenhouse is warmed by its transparent glass. From the end of the ice age (10,000 years ago) until 300 years ago, the atmosphere's level of CO_2 was about 260 parts per million (0.026%). At the beginning of the 20th century it was 290 parts per million, and today it is 380, heading for an estimated 400 to 550 in 2030 if current trends prevail.

The CO_2 in the atmosphere can come from being released from the oceans, from soils, and from the metabolism of microbes, plants and animals, but the recent rise in CO_2 is clearly due to human burning of fossil fuels, greatly increased since World War II. The decrease in temperature from 1940–70 seems to be correlated to increased burning of high-sulfur coal and oil, since sulfur dioxide (SO_2) and sulfate (SO_4) particles decrease the greenhouse effect. The 1991 eruption of the Philippine volcano Mt. Pinatubo released massive amounts of sulfur compounds, causing a brief cooling of the atmosphere.

The rise in *average* world temperature does not mean that all local regions are extremely warm, and there may be occasional colder than normal winters. Computer simulations of world climate predict increased heat waves, wildfires, droughts, stronger storm systems and hurricanes, melting of polar ice and mountain glaciers, and rising sea level. All of these have happened so far in the last couple decades. A sea level rise of eight inches has already been measured since 1900, and within a century, if the Greenland ice cap melts, sea level may rise several feet, drowning many low-lying coasts and cities. Climatologists estimate that if CO_2 doubles by the end of this century, average temperatures will rise by 3° to 8° F, compared to a rise of 1.1° F since 1900. Many wild species of plants and animals have already been found to have extended their ranges northward in the northern hemisphere, and the appearance of spring activities is now about two weeks earlier than 50 years ago.

One of the greatest worries of scientists is that ocean warming and the pouring of fresh water from melting ice into the oceans will quickly change the paths of river-like ocean currents, which will lead to drastic changes in wind and weather over the continents, including a rapid cooling of northern Europe (which is presently warmed by the Gulf Stream).

Future climate change will greatly impact world agriculture, with droughts, heat waves, increased need for irrigation, severe damaging storms, crops not able to grow in accustomed places, and increased food needs by a burgeoning human population. Switching a majority of agriculture to sustainable, biological or organic methods could go a long way to mitigating the problems, since these methods grow crops well-adapted to local conditions, which grow well in drought conditions, without irrigation, which need few if any inputs of synthetic fertilizers and pesticides, and which are more disease-resistant and nutritious than conventionally grown crops. Soil farmed organically has been found to "soak up" more CO_2 from the air than conventionally farmed soil, according to a study by the Rodale Institute. Converting 10,000 medium-sized farms to organic methods would reduce CO_2 as much as taking 1,174,400 cars off the road.

Unfortunately, some of the world's governments and politicians have not responded to the threat of global climate change as quickly or forcefully as necessary to prevent serious problems. Scientists say that even if human production of greenhouse gases was totally eliminated now, the natural processes already in action will cause temperatures to rise at least until 2050. Some experts say that if nothing significant is done within 10 years to combat global climate change, it may be too late to prevent worldwide catastrophe. Each person can do something in his or her own life to help. Reduce electrical power and fossil fuel use. Buy recycled products. It requires a change in thinking. Be less of a consumer; re-use "disposable" products. Urge local and national bureaucrats and politicians to do more. Everything positive that is done now will help, and possibly save modern civilization from collapse.

Problems and causes. As an agricultural example, a deficiency of the mineral element zinc in the soil will result in the plants growing in that soil being deficient in zinc. Zinc is a necessary part of certain plant enzymes, called dehydrogenases, which are vital in the metabolic functions of cellular respiration. So a zinc-deficient plant will have impaired cell

Wind damage to grain — perhaps the stems are too weak.

functions—it will not be a normal, healthy plant. Animals or humans who eat zinc-deficient food will suffer a variety of health problems, depending on conditions. Zinc helps heal wounds and skin irritations. Extreme deficiency can lead to delay in intestinal absorption of carbohydrates and proteins (poor utilization of food), low production of proteins and nucleic acids, skin problems, falling hair, sexual dysfunction, birth defects, and retarded growth. Zinc is a component of insulin, which is necessary for carbohydrate metabolism, and it can counteract lead and cadmium poisoning. Zinc is necessary in an enzyme called carbonic anhydrase, which functions in maintaining equilibrium between carbon dioxide and carbonic acid in tissues, which is important in the blood's carrying of carbon dioxide. Zinc is also necessary in digestive enzymes produced by the pancreas.

What might be the ultimate cause of the zinc deficiency in the soil? There may be plenty of zinc in the mineral materials making up the soil, but it may not be *available* to the plant. For example, at high (alkaline) pH, soil zinc is less available than at low (acid) pH. Also, zinc is not always just absorbed by plants as a simple ion. Zinc, along with several other elements,

Stunted corn (large area of field) in waterlogged soil.

can be more readily absorbed into plants in *chelated form*, an inorganic ion held and carried by a larger organic molecule. Natural chelators include certain organic acids, called humic acids, found in humus. Microorganisms are necessary to produce adequate humus in soil.

So what is the *cause* of sick animals or people in this hypothetical example? Is it the zinc-deficient food they ate? Or the low zinc availability in the soil the plants grew in? No, these are just symptoms. The real cause is something wrong in the soil. Perhaps toxic agricultural chemicals killed the microorganisms which produce humus. Or perhaps wrong fertilizers have altered the pH or "burned up" the humus. Or perhaps not enough manure and plant residues were returned to the soil. Whatever the cause or combination of causes, they can be eliminated or remedied, which will lead to adequate zinc availability, which will lead to healthy plants, which will lead to healthy animals and people. *How simple, once we understand the biological system.*

A concise slogan used by ecologists to explain what their science is all about is, "Everything is connected to everything else." This beautifully describes how natural systems are organized and function. A change in one part of the system affects the whole system to some degree.

Now that we have climbed our mountain of ecology and taken a quick look around at the landscape, let's go down the mountainside and plunge into the forest and examine the trees. Let's journey through the food chain and look at the parts and processes in more detail. First stop—basement!

Chapter 3

The Foundation —

SOIL

M y, it's dark down here in the soil. No wonder most people know so little about it. But that's why we're here, so let's learn. Soil is the absolute basis of agriculture, and thus of all human existence, for as we have seen, we either eat plants grown in soil, or animals which eat plants grown in soil. Our soil has been called our most important national resource. Wise use and management of the relatively thin upper layer, the topsoil, is vital for maintaining good health and a high standard of living.

But through misuse, about 7–10 *tons* of topsoil per acre are being lost to erosion each year in the Midwest (the figure can be much higher in the worst areas). It may take several hundred years for one inch of soil to form. Obviously, we can't keep on sending our topsoil down the river much longer.

Besides that, most once-fertile soil is now polluted by toxic substances, along with the groundwater and our wells. We are literally fouling our own nest, destroying the hand (or land) that feeds us! We had better understand more about this important, but neglected and abused part of the food chain. Let's take a tour through this dark and mysterious land that lies beneath our feet.

Parts. What is soil? It is a very complex substance, not "just dirt." Soil is a mixture of several components, sometimes defined as that part of the earth's surface capable of supporting plant life. It originally formed by the weathering of the rocks of the earth's crust. Typical soil contains approximately the following proportions of four constituents:

1. *Minerals* (about 45%), some of which are insoluble and not used by plants (sand, clay, iron oxides) and others which are soluble and provide valuable plant nutrients (calcium, potassium, magnesium). Mineral particles range in size from coarser gravel and sand to finer silt and the smallest clay particles, which are in the size range called colloidal.

2. *Water* (about 25%), which is needed as a part of plant cells and to dissolve and carry nutrients. Too much water in soil can exclude needed air.

3. *Air* (about 25%), which provides oxygen to roots and soil microorganisms, and nitrogen to nitrogen-fixing bacteria. Good aeration is vital to fertile soil.

4. *Organic matter* (about 1–5%), which includes the living soil organisms and the dead organic matter which decomposes to form humus. Humus has been broken down to very small particles in the colloidal size range. A "good" soil should have 2–5% organic matter, and up to 10% can be beneficial.

Colloids. In the above four components of soil, two contain colloidal particles, clay and humus. Colloidal particles, which are particles less than 0.002 mm ($1/5000$ inch) in size, are important in soil because they have a great ability to hold certain plant nutrients. Humus colloids can hold three times the nutrients that clay can. This is one of the reasons humus is so valuable in soil.

Tilth. A "good" soil should have a loose, almost spongy texture because the tiny soil particles (sand, silt, clay) are clustered into small clumps or "crumbs," also called aggregates. This condition is called good soil structure, or good tilth. Soil structure affects the ease of water penetration and aeration, root growth, the activity of soil organisms, and the availability of nutrients. Good tilth is generally only found in the upper layers of soil, with the lower, harder layer often being called a "hardpan," "plowpan," or "claypan." Soil of good tilth is easy to plow, soaks up water like a sponge, and resists erosion.

Factors that contribute to good soil structure include freezing and thawing, wetting and drying, penetration by plant roots, animal burrows, soil colloids, and most important, a "glue" secreted by roots and soil microorganisms. This is one of the many reasons soil organisms are important.

In poor soils, too much tillage can destroy tilth, as can leaving the soil without a vegetative cover to cushion raindrops and to encourage microorganism growth. Also, too acid or too alkaline conditions can destroy good tilth.

pH. One soil condition that agricultural "experts" concern themselves about excessively is pH. This is a measure of acidity or alkalinity of any substance using a scale of numbers, from 0 (most acid) to 14 (most alkaline), with 7 being neutral. The pH of some common materials includes: lemon juice = 2, vinegar = 2.5, black coffee = 5, pure water = 7, sodium bicarbonate solution = 8.2, ammonia water =11. Soil pH generally varies from 4 to 10, but most crops do best under slightly acid conditions (6.0–6.8). Soil pH

Roots have trouble penetrating a hardpan.

affects the availability of nutrients, which may be connected with nutrient deficiencies and toxicities (for example, manganese can be toxic to plants at pH 4.5 or below). Soil pH also affects the types of soil organisms and their ability to flourish, including those that fix nitrogen. Most bacteria cannot live in very acid conditions, while many fungi can.

The pH Scale

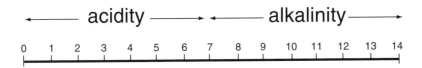

The traditional belief is that soil acidity is bad and should be counteracted by applying lime. But some acidity is necessary for plants to absorb certain nutrients from soil colloids. The experts usually base their liming recommendations on a single soil pH test, but testing several times during a year will reveal that pH can change markedly through a growing season, even daily. Also, various fertilizers and soil conditioners can have short-term and long-term effects on pH. So, as long as the pH does not become extremely acid or alkaline, pH "correction" and testing for many soils are not as important as some seem to think. The use of lime to "sweeten" soil is a case of doing the right thing for the wrong reason. Crops benefit more from the calcium supplied than from pH control.

Water. The movement of water in soils is little known or often misunderstood by most people. The average person assumes that water simply moves downward in soil, but if water is applied to soil at a single point, it defies the law of the gravity and moves just as fast horizontally as vertically; thus it soaks into the soil in a spherical pattern. Of course, rain normally falls all over the surface, not at a single point, so generally a horizontal "front" of water will soak in.

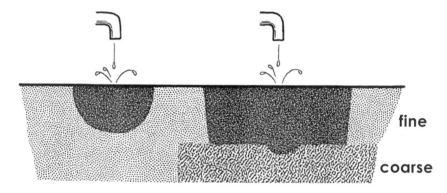

Left: water from a point source penetrates uniform textured soil in all directions; Right: a layer of coarser soil forms a barrier and retards water penetration.

One of the most amazing facts of water penetration is the barrier effect of a soil of different texture, such as a subsoil "hardpan" or simply a change in texture, such as from fine to coarse. When water encounters such a change in texture, its rate of penetration will be considerably slowed, and in reverse, a later rise of water to supply roots will also be slowed by such a barrier.

The role of porous organic matter, such as plant residue, in absorbing water can be important under the right conditions. Organic matter which is thoroughly incorporated into the upper layers of soil or which is partly exposed to the surface acts as a wick to increase water penetration. A compact plowed-under mass of organic matter acts as a barrier.

Organic matter. The importance of organic matter in soil cannot be overemphasized. William A. Albrecht, former head of the Soils Department at the University of Missouri, once called organic matter the "constitution" of the soil. As mentioned previously, soil organic matter consists of the dead decomposed humus and the living soil organisms. First let's look at humus, which has been called "the most important source of human wealth on this planet" (S. Waksman, *Humus*, 1938, p. 414).

Humus is a structureless colloidal material resulting from the decomposition (humification) of any type of dead organic matter (mostly plant residues and manures). It is a complex chemical mixture including proteins, lignin (originally part of plant cell walls), fats, carbohydrates, and

Common Types of Soil Microorganisms

Soil Bacteria

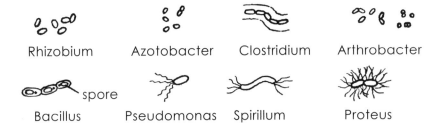

Rhizobium   Azotobacter   Clostridium   Arthrobacter

Bacillus   Pseudomonas   Spirillum   Proteus

Actinomycetes

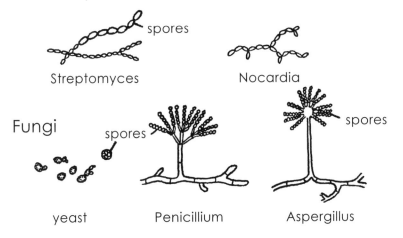

Streptomyces   Nocardia

Fungi

yeast   Penicillium   Aspergillus

organic acids. Its great ability to hold nutrients because of its colloidal nature has already been mentioned. The beneficial aspects of humus include:

1. It provides a storehouse of essential plant nutrients; for example, it stores over 95% of the nitrogen, 60% of the phosphorus, and 98% of the sulfur available to plants.

2. It helps make some nutrients more soluble and available to plants. Because of increased microorganismic activity during the warmer months, nutrients are released at the time of the plants' greatest need. Acids in humus also slowly dissolve soil minerals and release nutrients.

3. It contains substances that stimulate plant growth and improve crop quality and resistance to pests and diseases.

4. It provides a high water absorption and holding capacity because of its spongy nature.

5. It contributes to good soil structure (good tilth) by helping make soil crumbly and porous. It also reduces wind and water erosion and makes soil easier to work.

6. It buffers the soil and protects plants from high salt levels, toxic chemicals, and drastic changes in pH.

7. It provides food for the beneficial soil organisms, especially in the "pre-humus" form of fresh organic matter.

So we see that humus is a marvelous substance, with many functions, and "worth its weight in gold." Now let's look at its equally important partner, the living part of the soil.

Fertile soil is literally teeming with an amazing variety of plant and animal life, most of which are microscopic and thus little known nor appreciated by the average person. They form a valuable "work force," performing a multitude of chemical transformations, as well as many other services. An idea of their abundance can be gained from the following table, which shows figures for a temperate grassland soil (from L.M. Thompson & F. Troeh, *Soils and Soil Fertility*, 4th ed., 1978):

organism	number/acre	lbs/acre
bacteria	800,000,000,000,000,000	2600
actinomycetes	20,000,000,000,000,000	1300
fungi	200,000,000,000,000	2600
algae	4,000,000,000	90
protozoa	2,000,000,000,000	90
nematodes	80,000,000	45
earthworms	40,000	445
insects & other arthropods	8,160,000	830

According to one estimate, one teaspoon of rich grassland soil could have in it 5 billion bacteria, 20 million fungi and 1 million protozoa and algae. That's nearly as many microbes as the world's human population.

Nitrogen Fixing

The ability to transform the chemically stable gas nitrogen, into organic nitrogen compounds is found in only a few groups of microorganisms: about 20 types of bacteria, 10 types of blue-green algae, also called cyanobacteria, two types of yeasts, and probably one type of actinomycete. The most important to agriculture are the bacteria, which include both aerobic and anaerobic species, most of which are common in the environment and live in soil independently of plants. The best known of these free-living bacteria are the aerobic *Azotobacter* and the anaerobic *Clostridium*. These bacteria grow best when there are carbohydrates, which can come from organic matter (manures and crop residues) added to the soil as a source of food. "When large quantities of carbohydrates are incorporated into soil, significant increases in combined nitrogen occur" (T.D. Brock, *Biology of Microorganisms*, 1970, p. 578). In laboratory test jars of composting dairy cow manure, I have found an increase in nitrate nitrogen of over 60 times due to the action of nitrogen-fixing bacteria.

The best known nitrogen fixing microorganism is *Rhizobium*, the type of symbiotic bacteria which grows in the root nodules of legumes, such as alfalfa, clover, beans and peas. Neither the legumes nor the bacteria alone can fix nitrogen, only when they "cooperate" can this feat be done. The formation of root nodules is amazing. The *Rhizobium* bacteria are among the common rhizosphere organisms that grow on plant roots. One of the materials secreted by legume roots is the amino acid tryptophan. *Rhizobium* changes tryptophan into the plant hormone indoleacetic acid, which causes the root hairs to curl. Then some of the *Rhizobium* change into highly mobile "swarmer cells" which work their way through the root hair's cell wall. Once inside, the swarmer cells grow into a thread-like structure that invades cells of the root's cortex region. If a root cell happens to have twice its normal number of

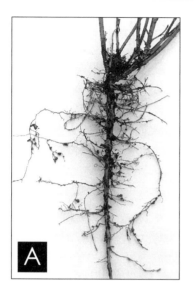
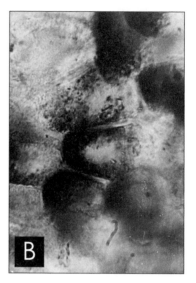
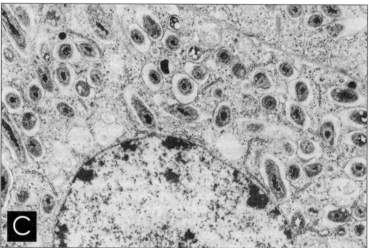

A. Root nodules of alfalfa; B. nitrogen-fixing bacteria inside alfalfa root nodule cells, magnified 100 times; C. nitrogen-fixing bacteria (smaller round and sausage-shaped objects) inside a nodule cell (large rounded structure at bottom is the nodule cell's nucleus); magnified 5,000 times (photo courtesy Dr. Shiv Tandon).

chromosomes (a small number of such cells are always present), it is stimulated to begin dividing and producing a nodule. The bacteria then rapidly multiply and fix nitrogen inside the inner nodule cells. A "healthy" nodule will have a pink or red color inside because of a red protein called leghemoglobin, which is similar to the red hemoglobin that carries oxygen in our blood.

The *Rhizobium* bacteria can survive in the soil for many years, waiting to reinfect new legumes, IF soil conditions are favorable. Needless to say, anything farmers add to their soil which kills soil life cheats them out of about 80 pounds per acre of free nitrogen fertilizer.

Several types of plants other than legumes can have nitrogen fixing nodules that contain microbes other than *Rhizobium* (blue-green algae and a probable actinomycete). Research is currently being done to try to get non-legume crops to grow nodules.

The first four types of organisms in the previous table (p. 42) are plants and the last four are animals. The most important are the bacteria, actinomycetes, and fungi. Bacteria are single-celled organisms that can live in either aerobic (aerated, with oxygen) or anaerobic (without oxygen) conditions, but generally cannot live in very acid soils. The aerobic bacteria are the beneficial kind in general. Some of the beneficial activities of soil bacteria include helping decompose organic matter to form humus, converting inorganic chemicals into useful plant nutrients, detoxifying (breaking down) man-made toxic chemicals (herbicides, pesticides), and fixing (trapping) nitrogen from the air for later plant use.

Actinomycetes are filamentous (thread-like) organisms that are considered to be intermediate between bacteria and fungi. They require aerobic, neutral or slightly alkaline soils. They aid in producing humus.

Fungi, or molds, are filamentous, mainly aerobic, and can tolerate fairly acid conditions. Besides those helping to produce humus, one particular group of fungi, the mycorrhizae, are especially important to plants. They are symbiotic and live partly inside the roots of many kinds of plants. They help roots absorb nutrients (especially phosphorus and nitrogen), secrete growth-promoting hormones, and help protect roots from disease organisms.

Other important functions of soil organisms besides those already mentioned include temporarily "tying up" nutrients from minerals and organic matter in their own bodies, as part of the proteins, carbohydrates, and fats of their cells. By doing so, they counteract the loss of some nutrients from the soil by leaching into the groundwater. When the organisms die, the nutrients will be made available to the plants, which are thus fed slowly over the growing season. Another important function of soil organisms is, along with humus, promoting good soil structure. The "glue" that soil microorganisms produce, which is a type of sticky carbohydrate, has been previously mentioned. It holds soil particles together, forming "good crumb structure:" loose, crumbly, porous soil.

Mycorrhizae

The word mycorrhiza literally means "root-fungus." Mycorrhizae are special symbiotic fungi that live on or in the roots of plants. At first, scientists thought that they were a disease and that they were found on only a few kinds of plants, but now it is known that nearly all (80%) plants can have mycorrhizae and that they are a very beneficial kind of fungus, not a disease.

There are two kinds of mycorrhizal fungi. The ectomycorrhizae grow around the outside of the plant's roots and between root cells, but not inside them. They are mainly found on trees. Roots that are "infected" with ectomycorrhizae have a "fatter" appearance than non-infected roots. Some of the species of mushrooms which always grow around certain trees, both poisonous and edible ones (including the gourmet's delight, truffles), are actually the fruiting bodies of ectomycorrhizae. The other kind, called endomycorrhizae, occur on plants other than trees, and actually grow inside some of the cells of the plant's roots. Just one ounce of grassland soil can have as much as 1.75 miles of mycorrhizae.

Mycorrhizae perform a number of valuable services for the plant with which they are associated.

1. They increase the plant's absorption of nutrients, especially phosphorus and nitrogen, and also immobile elements such as zinc, copper, and molybdenum, and the ammonium ion. In a 1937 study of white pine, mycorrhiza-infected trees absorbed 1.24

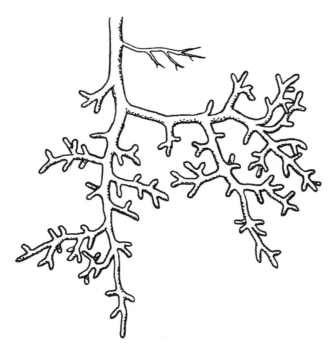

Tree roots with mycorrhizae have a stumpy appearance, while non-mycorrhizal roots are slender (top center).

mg nitrogen, 0.196 mg phosphorus, and 0.744 mg potassium (% of dry weight), compared to non-mycorrhizal tree figures of 0.85 mg nitrogen, 0.074 mg phosphorus, and 0.425 mg potassium (J.L. Harley, *Mycorrhiza*, 1971). Nutrient absorption can be increased partly by the fact that the fungal threads increase the absorptive area of the root and can reach out farther than the plant's root hairs. Also, mycorrhizae aid in nutrient absorption by utilizing nutrients tied up in soil organic matter. Thus mycorrhizae can change some nutrients into forms useable by plants.

2. Mycorrhizae produce hormones that stimulate the growth of their host plant. This plus the increased nutrient absorption just mentioned can combine to increase plant growth significantly—sometimes by several hundred percent.

3. Mycorrhizae can increase water absorption by a plant because they can grow at lower soil water levels than plants.

4. Mycorrhizae can protect roots from diseases, either by forming a protective barrier or by producing antibiotics.

In return for its services, the fungus receives a supply of carbohydrates and certain other nutrients from its host plant; thus both partners in the symbiotic relationship benefit. Mycorrhizae benefit plants most when the plants are growing in poor soil, particularly phosphorus-deficient soil. Plants growing in fertile soil do as well with or without mycorrhizae.

Even though nearly all crops can have these helpful mycorrhizae (cabbage, spinach, and garden beets are exceptions), many farmers are missing out on their valuable services because their soil is not hospitable for their existence. Mycorrhizae, as well as other beneficial soil microorganisms, need the following conditions:
1. adequate moisture
2. oxygen
3. proper temperature
4. proper pH, not too acid or too alkaline
5. adequate soil organic matter
6. freedom from fungicides or other soil sterilants
7. a low concentration of salts, including soluble fertilizer salts (especially nitrates)

Chemical Warfare in the Soil

Few people are aware of the complex interactions among the many species of microorganisms inhabiting soil, and the interactions between them and plant roots.

Besides the dozens of species of bacteria, actinomycetes, and fungi that live on the soil particles away from plant roots, there are other species which only live in the film of sticky material secreted by plant roots (root exudate). The region close to roots is called the rhizosphere, and the rhizosphere microorganisms play an important role in the well-being of the plant [the special mycorrhiza fungi and the nematode-trapping fungi are covered in detail in other boxes, in this chapter and in Chapter 16].

Soil conditions are constantly changing, both on a large scale (temperature, moisture, pH, etc.) and on a microscopic scale; different conditions may exist a few feet or a fraction of an inch away. For example, if the pH is above 6, the nitrogen-fixing bacteria *Azotobacter* can flourish. Well aerated soil will allow aerobic bacteria to grow, as well as fungi and actinomycetes. But a heavy rain which accumulates in a low area will exclude air and allow harmful anaerobic denitrifying bacteria, such as *Pseudomonas*, to release nitrogen, while fungi, actinomycetes, and aerobic bacteria become temporarily dormant.

Besides the nitrogen taken from the air by nitrogen-fixing bacteria and algae, other soil microorganisms aid plants by helping to make soil nutrients available (partly by producing acids that "break out" mineral nutrients) and by producing various substances useful to plants—amino acids, growth hormones and vitamins.

But the most fascinating way in which soil microorganisms interact with plants and other microorganisms is by various chemicals that they produce and secrete by metabolic processes— a sort of chemical warfare. Certain common soil bacteria, actinomycetes, and fungi can produce substances that inhibit or kill other species of microorganisms, especially those that cause plant diseases. Sometimes they only do this when the disease-causing species are present. Such human-used antibiotics as penicillin, streptomycin, and tetracycline are grown from the soil fungus *Penicillium* and the actinomycete *Streptomyces*. The Russian researcher, N.A. Krasil'nikov, in his book, *Soil Microorganisms and Higher Plants* (1958, English edition 1961), states that there are natural microbial enemies of all species of soil microorganisms and that they occur by the hundreds of thousands or millions per gram of soil! (This is assuming that the soil is "alive," and not sterilized by wrong agricultural practices.) He states, "Through use of their metabolic products antagonistic microbes suppress their competitors, removing them from the substrate and thus exerting a definite selective action. To a certain degree, microbial antagonists regulate the formation of microbial coenoses (colonies) in the soil in general. They play an important role in the improvement of soils, in the so-called process of self-purification of soils. The removal of harmful pathogenic and phytopathogenic flora and fauna is accomplished by microbial antagonists. . . . Antagonists, therefore, can be considered as one of the powerful factors governing soil fertility and plant-crop abundance." (p. 364–365).

It has been found that the contents of plant cells (the "sap") have the ability to suppress the growth of bacteria and fungi. N.A. Krasil'nikov believes that organic substances produced by soil microorganisms, when absorbed through the plant's roots, increase this natural immunity of plants to infections. He finds that the most practical way to insure that the soil has these beneficial organisms is for it to be rich in humus and by fertilizing with manure or compost. More recent research has zeroed in on natural iron chelates produced by rhizosphere bacteria. The theory is that these chelates tie up much of the iron, making it unavailable to disease-causing microbes—sort of starving them out. Methods for large-scale application of growth-promoting and disease-suppressing microbes are being developed; seed inoculation has been the most successful so far. Yield increases of up to 30 to 144% have been reported.

The humus and soil life work together as a team and perform all these valuable services—and the best part is, they work for free, if only given half a chance! Why would anyone want to dump toxic materials on his land that would kill or inhibit this volunteer army? Are the experts warning us about this?

Soil quality—the problem of soil air. When you see a small lake in a field two days after a rain, or a stunted field of corn with yellowish leaves—do you wonder why—what went wrong? Or do you just shrug your shoulders and think, "Oh, that's the way all the fields are this year"?

We need to do more questioning, probing, and searching. We need to be aware of problems and then find out what's causing them. What's *really* causing them. Only then can we solve the problem—by attacking the real cause or causes, not just trying to cover up the symptoms as so many "experts" are doing these days.

Soil air. One of the worst and most widespread problems today is anaerobic soil. Now before you get scared off by that technical term, let's explain. *An-aerobic* simply means "without air," or "without oxygen." The opposite term is *aerobic*, meaning "with air" or "with oxygen." When these terms are applied to soil, they refer to the relative amount of air or oxygen in the pore spaces, or to well-aerated or poorly aerated soil. Previously we mentioned that typical soil contains about 25% air. This applies to topsoil, and it is a rough average. The pore space between solid soil particles occupies about 50% of the soil's volume (in good loose soil), and the pore space can be filled with either air or water. Usually the amount of water fluctuates between 15 and 35% (and the same for air; when the percent of water is higher, the percent of air is lower).

Sandy soils have large pores between sand grains and are usually well aerated. Clays have much smaller spaces between the fine clay particles and can easily become "tight" and poorly aerated. A soil with good crumb structure (good tilth; with "crumbs" or aggregates) will be well aerated.

In any soil, the upper layers of the soil will be aerobic and have more air than deeper soil. Soil air comes from the atmosphere, so the deeper you go in the soil, the less air there will be, and anaerobic conditions will prevail. Actually, on a microscopic scale, there is usually a mixture of aerobic and anaerobic conditions. The center of a soil crumb may be anaerobic, while a few thousandths of an inch away, near a pore, there may be plenty of oxygen.

Importance of air. Why is air or oxygen so important in the soil? There are a host of interconnected reasons, but there are two main things in the soil that need oxygen: plant roots and most of the beneficial soil organisms. Roots need oxygen to function properly and to live. More about that later.

The beneficial soil organisms (most types of bacteria, plus fungi, actinomycetes, algae, protozoa, earthworms, etc.) need oxygen to survive. They are called aerobic organisms, or aerobes. Some bacteria can live and grow without oxygen; they are called anaerobic organisms, or anaerobes. Some talented bacteria can switch back and forth depending on their environment's oxygen supply; they are called facultative anaerobes. Some of the anaerobic bacteria are beneficial (some decompose minerals to release nutrients; one type fixes nitrogen), but most are harmful to plant growth and the soil's nutrient and humus content. So, the better the soil is aerated (with good structure or tilth), the better it is for growing crops (in rare cases of over-aeration, humus will be "burned up" too quickly). We covered the beneficial functions of soil organisms and humus earlier in the chapter.

Signs. What are the signs of anaerobic soil? Soils that are "tight" and heavy are more apt to be anaerobic. Crusting and compaction are bad signs. Another common sign is standing water and waterlogging, which indicate tight soil (little pore space) or an underlying hardpan (or plowpan, claypan). Humus levels are probably low, and often the soil is practically sterile, without earthworms and other life. If you dig up a plant (corn is especially good) and shake off the soil, the depth to which small feeder roots extend is a good indication of how deep your soil's aerobic zone is. These signs of trouble are all too common in fields across the country today. Let's look at the problems anaerobic soil can cause.

Anaerobic problems. The number and variety of problems that can result from anaerobic soil are amazing. Some affect the plant directly and immediately, while others are delayed or indirect.

An anaerobic soil condition first of all means that the oxygen content is low and the carbon dioxide level is high. Well-aerated soil may have an oxygen content only slightly lower than in the atmosphere (21%), but in soil with poor structure or because of farming practices that decrease aeration, the oxygen level may be below 1%. Roots and soil organisms can deplete the available oxygen supply in a matter of hours.

The atmosphere only has 0.038% carbon dioxide. Normal soil has about 0.1–1.5% (non-agricultural soil) to 3% (cropped soil). Anaerobic and waterlogged soils can have up to 15–20% carbon dioxide. These levels are "directly toxic to plants . . . the capacity of roots to absorb water and to a lesser extent, ions [nutrients], is sharply reduced," according to R.D. Durbin in *Water Deficits and Plant Growth*, Vol. V, 1978.

Changes in soil chemistry. These large shifts in oxygen and carbon dioxide content can cause adverse conditions in the soil. If the soil is high in calcium, high carbon dioxide forms carbonic acid, lowering pH (more acidity). A 10% carbon dioxide level decreases pH two units (100 times, since each pH unit is ten times the one above or below it). In low calcium soils, pH can be raised instead. When pH is lowered especially, the availability of several elements changes and nutrient imbalances and toxicities can result. For example, aluminum and manganese become more available (and possibly toxic) at low pH, while nitrogen, phosphorus, potassium, magnesium, calcium and sulfur become less available.

At low oxygen levels there are other chemical changes causing toxicities. Certain forms of iron and sulfur (ferrous and sulfide ions) can build up to toxic levels. Also the availability of several trace elements and heavy metals is increased, possibly to toxic levels (these include boron, cobalt, molybdenum, nickel, lead and zinc).

Finally, under waterlogged conditions, certain anaerobic bacteria rapidly change nitrate nitrogen (which plants need) into nitrites (toxic to plants) and then into gaseous nitrogen or nitrogen oxides (which escape into the air). Close to 100 lbs./acre of nitrogen can be lost in a growing season due to denitrification.

Furthermore, waterlogging dissolves excess salts, either from the soil's minerals or from previous fertilizer applications. Roots are injured, plant growth is slowed, and excess salts are carried to the leaves where they can accumulate and kill areas of the leaves of sensitive species ("leaf scorch"). The dying leaf or root tissues allow fungal diseases to get started.

Upset soil life. In "healthy" soil, the many species of microscopic organisms live in a constantly fluctuating balance, with one species influencing another. Some help each other and some are "antagonists," keeping each other in check.

In waterlogged soils, "within a matter of hours the soil microflora changes dramatically, particularly in the rhizosphere. Fungi and actinomycetes almost disappear while many largely anaerobic bacterial species now flourish," said R.D. Durbin (1978).

The anaerobic bacteria carry on metabolic activities (feeding, excretion) sometimes called fermentation or putrification. They release a variety of chemical by-products which are toxic to roots in large enough amounts: hydrogen, ammonia, methane, hydrogen sulfide, carbon dioxide, and ethylene.

Upset plant functions. Anaerobic conditions and waterlogging cause drastic changes in the functioning of the entire plant, perhaps only a few at first. But one problem leads to another, and pretty soon the plant is hopelessly sick, growth slows, and yields are cut. We have already mentioned how low oxygen, high carbon dioxide, and altered soil chemistry can disturb or damage roots. Only a short period of anaerobic conditions will retard a plant's functions, even though it will recover and appear to be normal and healthy. A single day of anaerobic conditions (produced by flooding) can reduce the dry weight growth of tomatoes as much as 30%. Some crops such as soybeans, apples and grapes are more resistant, however.

Hormone changes. At least three plant hormones can be thrown out of balance by anaerobic conditions. Ethylene production is increased, and the transport of cytokinins and gibberellins from the root is decreased. This causes reduced stem growth and reduced chlorophyll production in the leaves. Sometimes abnormal growth of roots from the lower stem and of "tumors" on the stem will occur. The disease called gummosis can also be triggered (see next section).

Gummosis. When a plant is under stress, it may produce slimy, sticky, gummy materials in its tissues. They can plug up some of the tubes of the vascular system, the vessels that carry water and nutrients upward from the roots (called the xylem tissue) or the cells that carry food from the leaves to other parts of the plant (called the phloem tissue). This condition of "plugged plumbing" is called gummosis by plant disease specialists. Leaf wilting, reduced growth, and death of some cells or of the whole plant can result. Gummosis can be caused directly by anaerobic soil conditions or from toxins released by anaerobic bacteria, from unbalanced plant hormones, from disease or insect attack, and from soil nutrient deficiencies.

Fermentation. Because of oxygen deficiency in the soil, the plant's metabolism can be upset, with anaerobic respiration (fermentation) replacing the usual aerobic respiration. A by-product of fermentation is

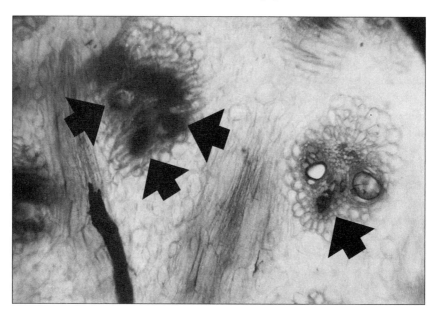

Vessels of corn plugged by gummosis. Magnified 100 times by microscope.

ethyl alcohol (as moonshiners and brewery workers know!). Alcohol levels in the plant's cells can build up to toxic levels. In one study, it took only 2 1/2 hours for potato tubers to become anaerobic when covered by a film of water, which reduced the rate of oxygen uptake (see *Potato Research*, Vol. 13, 1970).

Diseases and pests. Healthy plants can resist attack by disease pathogens ("germs") and pests (insects, mites, nematodes). Stresses can cause a plant to lose its natural immunity. The stresses resulting from anaerobic soil can allow diseases and pests to get a foothold. Certain microorganisms which are normally present in soil and do not attack healthy plants are definitely known to attack stressed plants under anaerobic conditions.

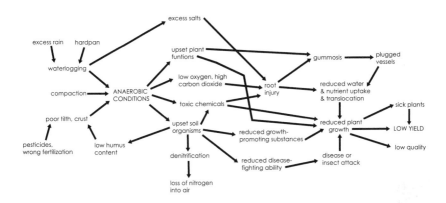

Overview. Whew! That's quite a list of problems resulting from poorly aerated or waterlogged soil. In general, all of these results of anaerobic soil conditions have the effect of disrupting the soil's beneficial organisms and the normal functions of the plant. Thus the growth and health of the plant suffer and yields and/or crop quality are reduced. They are sources of unnecessary stress on the plant. Other factors can cause stresses also: drought, high temperatures, cloudy weather, unbalanced soil fertility, etc. Any stress is bad for the plant and can lead to diseases, pests, and reduced yield and quality. Why add unnecessary stresses? The harmful effects of anaerobic soil can be overcome *and prevented*. Let's see how.

Causes. Anaerobic, tight and waterlogged soils are mainly caused by (1) undesirable soil types, with too much or too little of some component (clay for example, or salts, or alkalinity); (2) low humus content; (3) too much tillage and traffic; (4) wrong fertilization, such as anhydrous ammonia which in typical amounts kills soil life, causes drastic pH changes and dissolves humus; and (5) overuse of pesticides (herbicides, fungicides, insecticides and others), which may change soil structure and upset the balance of soil life.

Correcting anaerobic soil. Now that we have seen what causes anaerobic soil, we can then look at the other side of the coin and see how to correct problem soil and how to prevent soil problems. These corrective methods solve a multitude of problems.

1. *Increase humus and soil life.* If soil is not in extremely poor condition, work fresh organic matter (manure, crop residues, green manure, sewage sludge or other by-products) into the upper several inches of soil. This should be done in the fall if possible, or else a small amount a few weeks before planting a crop. The mixing of both plant and animal wastes gives better results, and the mixing of rock fertilizers (lime and soft rock phosphate; greensand if potassium is needed) gives even better results. Do not work fresh organic matter deeper in soil than the present aerobic zone or a little deeper (look at the depth of plant feeder roots to tell how deep the aerobic zone is).

If soil is very poor, work in compost rather than fresh organic matter. A little poultry manure added will provide an abundance of beneficial bacteria. Rock fertilizers mixed in will increase the benefit.

Leaving a field in a crop of a legume (clover, alfalfa, vetch, etc.) or especially perennial grasses (smooth brome, fescue, bluegrass, timothy, reed canary grass, etc.) for a year or more is one of the best ways to increase humus and improve soil structure (because of all the small roots).

Any of those things, cover crops, fresh organic matter, compost, and rock fertilizers (high calcium lime and soft rock phosphate especially), will help loosen tight soil, prevent crusting, reduce erosion, and increase water retention. A good earthworm population will break up a hardpan.

2. *Use temporary helps.* Until soil tilth improves, cultivate to aerate upper soil layers (cultivation is also important for weed control). Subsoiling will break up a hardpan, at least temporarily. Tilling a little deeper each year will gradually eliminate a hardpan, especially if humus is increased at the same time. Some commercial soil conditioners (wetting agents or surfactants, humic substances or humates) can help loosen soil and break up a hardpan, at least sometimes. They can be expensive, but may be a temporary help. Products that inoculate soil with algae or bacteria sometimes work well, but again may be expensive. Ground-up rock

fertilizers (lime, rock phosphate) generally help improve soil structure, and seaweed products can also have beneficial effects.

3. *Reduce compaction.* Use lighter machinery, avoid unnecessary trips over the field, and do not work soil that is too wet.

4. *Reduce tilth-destroying chemicals.* Reduce or eliminate the fertilizers and pesticides that tend to destroy soil structure, humus, and/or life.

For the next stop on our journey, let's take the elevator one floor up and wander about in the green jungle of the plant world.

Chapter 4

Food Factory —

PLANTS

H ere we are at ground level, walking around in the wonderful sunlight. Makes you feel like stretching your arms out and soaking up some sun. That's it, go ahead; you're already starting to think like a plant!

Plants are such different organisms from ourselves and the animals that we are more familiar with, that they seem to live in a different world. They have no eyes or ears, no bones, nerves or muscles. They can't walk or run around from place to place. But believe it or not, plants are sensitive to light and vibrations, have a supporting skeleton, exhibit electrical activities, and do considerable moving. They are living, "breathing," growing organisms, and anyone who makes a considerable part of his living growing them should know more about them.

To understand more about how plants function, let's shrink ourselves down to the size of a water molecule (only about 0.000006 inch) and take a trip through a plant. A living plant is very much like a bustling city, only much more complex. It is built up of thousands of units called cells, as a city is largely composed of buildings. Some are factories, some warehouses, some power plants, some houses. Similarly, different cells have different functions: some are "factories," some for storage, and some form a transportation system, like the plumbing of a city. The people and workers in the city would correspond to the many kinds of active ions and molecules, enzymes and energy carriers that help "run" a living plant. A plant is incredibly complex, so we will only hit the highlights, especially those that relate to agriculture.

To begin our trip through a living plant, we will have to once more venture down into that murky realm of the soil, which we recently visited, but this time we will take a different perspective. If we were the size of a water molecule (0.000006 inch), we would find the soil to be an awesome and bizarre place indeed. We would probably be oozing along in a film of water on the surface of a huge, platelike colloidal clay particle, a gigantic 0.005 inch across. Everything would be in motion, for atoms, ions, and molecules of all matter are constantly vibrating and floating around. We would see fuzzy-looking spheres and lumps of various sizes floating and zooming by. These would be ions and molecules of water, hydrogen, carbonate, calcium, potassium, ammonium, nitrate, and so on. [If you are not familiar with these chemical terms, consult the Glossary and Appendix A.]

We would also have to duck out of the way to avoid being devoured by immense (0.00002 to 0.004 inch) monsters snuffling and gurgling about, such as bacteria and protozoans (one-celled animals), or even an incredibly gigantic, snakelike nematode, or threadworm, 0.04 inch long.

Such is the world of the rhizosphere, the zone close to a plant's roots. Plants primarily absorb two things from the soil—water and nutrients. If a plant doesn't have enough of either, the farmer may as well start over, because he won't get a crop. So we should know a little about how these substances are taken up by plants.

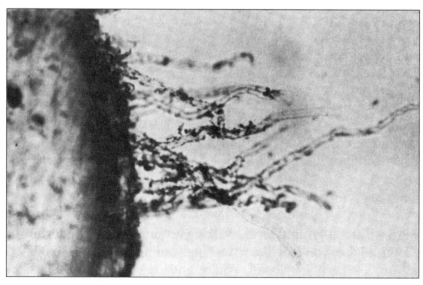

Root hairs of corn, magnified 100 times under the microscope. Note the tiny dark soil particles stuck to some of the root hairs.

In 1937, H.J. Dittmer measured the roots of a single rye plant. He found that the main and smaller roots had a total length of over two million feet (387 miles) and had a surface area of 2,554 square feet, while the tiny root hairs attached to the smaller roots had a total length of 6,604 feet (1.25 miles), but a surface area of 4,321 square feet. Altogether, root and root hairs had a surface area of over one-sixth of an acre!

Water. The entrance of water into plants is controlled by osmosis. Osmosis involves the relative percent of water on the outside compared to the inside of a cell. Water moves (diffuses) from wherever it is more concentrated (higher %) to where it is less concentrated (lower %). Thus if there is more water in the rhizosphere than inside a root, it will easily and automatically "seep" (diffuse) inside. Plants have a marvelous set of tiny root hairs at the tip of each root to add surface area for better absorption. The only complicating factor is that some water molecules are so tightly held on the surface of colloidal soil particles that they can never be absorbed by roots.

Every farmer and gardener has experienced the fact that when plants do not absorb enough water, they wilt and die. But is this always caused by drought—not enough water in soil? That can be one cause, but the same result can happen when there are too many salts dissolved in the soil water, lowering the *relative percentage* of water outside the root cells. When there is a higher percentage of water inside the cells than outside, reverse osmosis will occur; the roots will lose water rather than absorb it, and root damage or death of the plant will result. Salt damage is not just a problem in the alkaline and saline soils of the western U.S.; it can happen anywhere if a farmer applies too much salt fertilizer or manure to the soil (manure contains salt from the animal's diet). Plenty of rain can alleviate the situation by diluting or leaching excess salts.

One other subject relating to water use by plants is humus. Remember one of the functions of humus listed in Chapter 3 was absorbing and holding water, like a sponge. A soil with plenty of humus can produce beautiful crops even in a drought year.

Back to our trip through the plant. Once water reaches the central "core" of a root, where the "plumbing" is (the vascular tissue, xylem and phloem), it enters into the water-carrying tissue, the xylem, for a quick elevator ride up the roots and stem to wherever water is needed in the plant.

Nutrients. But wait a minute. The roots absorb two materials, water AND nutrients. We next must learn something about nutrient uptake. Here again we get into the complications of soil chemistry. Plants need a variety of chemical nutrients. The essential plant nutrients can be divided into the macronutrients, needed in larger amounts, which are nitrogen, calcium, phosphorus, potassium, magnesium, and sulfur; and the micronutrients, or trace elements, *just as important*, but needed in very small amounts. The micronutrients are iron, copper, zinc, cobalt, boron, manganese, molybdenum, and chlorine.

These nutrients can be taken into a plant in various forms, some as simple ions (atoms with a positive or negative electrical charge, such as potassium, calcium and zinc), some as molecular ions (the nutrient atom combined with others, such as nitrate, ammonium, and sulfate), as chelates (the nutrient atom held by a larger organic molecule, such as iron, zinc, copper, manganese, and cobalt chelates), or as organic molecules that are precursors of proteins or other cell components (such as phenol, amino acids, amines and indole). Conventional agriculture as commonly taught in schools today virtually ignores the latter two, and assumes that a plant can absorb only inorganic ions and still produce quality crops.

But back to our story of nutrient uptake. The inorganic ions that carry a positive charge are involved in an interesting phenomenon called base exchange, or sometimes cation exchange (a positively charged ion is called a cation, a negatively charged ion is called an anion). The clay and humus colloidal particles carry a negative charge on their surface, which attracts and loosely holds positively charged ions, some of which are plant nutrients. Since they are loosely held, they can easily trade places with another positive ion in the vicinity. This is base exchange. One of the by-products of the metabolism of plant roots and soil microorganisms is hydrogen ions, which have a positive charge, and which can be exchanged for nutrient ions. The more active a plant is, the more nutrients it needs and the more hydrogen ions its roots produce, and the more nutrients it can absorb by base exchange. Isn't that a beautiful, self-regulating system for supplying just the right amount of nutrients at the right time? If there's a way to foul up something, someone will do it. Too heavy an application of the highly soluble salt fertilizers will "flood" the soil water with so many ions that the plant will be forced to take in more than it needs. This can upset the plant's metabolism and in some cases be toxic to the animals

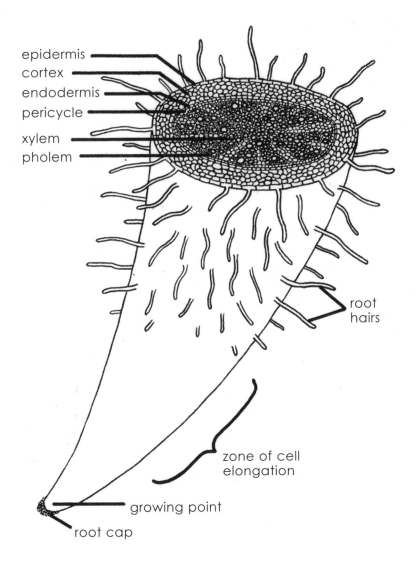

epidermis
cortex
endodermis
pericycle
xylem
pholem

root hairs

zone of cell elongation

growing point

root cap

A root is composed of several parts. The root cap protects the growing point. The many root hairs add absorptive surface area. The epidermis covers the exterior. Internally, the cortex stores food, the endodermis regulates inward passage of absorbed materials, the pericycle produces branch roots and additional xylem and phloem, the xylem carries water and minerals upward, and the phloem carries food downward.

that eat it (an excellent example being nitrate poisoning from animal feed grown with too much nitrogen fertilizer). We also see from base exchange another reason for the value of soil organisms—releasing hydrogen ions.

Caution, Roots at Work!

We all have some appreciation of the work that the above-ground parts of plants do, especially the leaves making food by photosynthesis, as well as flowers producing grain or fruit. But because they are hidden underground, few people realize the importance and wide range of activities that roots perform. In fact, savvy agronomists say that a good root system is more important than the hybrid variety, fertilizer or herbicide in determining final yield. Roots and their activities are amazingly complex.

In overview, roots 1. *anchor the plant*, giving lodging resistance; 2. *absorb water*, and if healthy, provide drought resistance; 3. *absorb nutrients*, especially important for final yield and crop quality; 4. *hold soil*, reducing erosion; and 5. *improve soil*, by adding organic matter, producing good soil structure and sometimes adding nitrogen.

The rhizosphere. Most of the activities of roots take place in the rather narrow (about 1/16 to 1/8 inch) region immediately around all root surfaces, called the *rhizosphere*. Within the rhizosphere, conditions are much different than in the surrounding soil.

Near their tips, roots produce a watery, slimy material, called *exudate*, which seeps into the rhizosphere. The root exudate contains water, along with a host of substances: carbohydrates (sugars and polysaccharides), amino acids (about 20 different ones), organic acids (such as acetic acid, citric acid, oxalic acid, propionic acid, etc.), fatty acids and sterols, vitamins (mainly B vitamins), plant hormones and growth regulators, enzymes, and several other types of substances, including phenolics, nucleotides and organic phosphorus compounds.

The root exudates do several things: 1. aid the growing root tip to push through the soil; 2. moisten the soil to aid in absorption of some nutrients, such as phosphorus, potassium, zinc, copper, iron and manganese; 3. help control soil pH; 4. feed and stimulate the growth of beneficial soil organisms, such as bacteria, actinomycetes and fungi, including those very important nitrogen-fixing bacteria and mycorrhizal fungi (see the boxes in Chapter 3); 5. improve soil structure (create soil aggregates) by gluing soil particles together and stimulating soil organisms that do the same; 6. fight harmful organisms (see box in Chapter 3, "Chemical Warfare in the Soil"); and 7. sometimes harm other nearby plants by releasing toxic substances, a process called *allelopathy* (this is one way weeds can harm crops; however, some crops can inhibit weeds, including small grains, sorghum, corn, soybean, buckwheat, alfalfa and red clover).

Health. interestingly, a plant's state of health greatly affects the vital functions of roots, root exudates and the beneficial organisms of the rhizosphere. For example, if a plant is under stress, perhaps from drought or low-oxygen soil, changes in just one or two amino acids in the exudate may stimulate spores of disease microbes to germinate and attack the plant.

Besides weather conditions, agricultural practices may greatly affect the health of crop root systems and thus the growth and productivity of the crop. Throughout much of the rest of this book, we will explore good and bad aspects of fertilization, liming, weed and pest control, tillage, organic matter recycling, and so on.

Chelates

The word *chelate* comes from the Greek word *chela*, or claw, referring to the holding ability of these chemicals. Chelates are organic molecules that are able to hold and release certain metal ions, including such plant nutrient elements as calcium, iron, magnesium, cobalt, copper, zinc, and manganese. These elements are more easily absorbed by plant roots in chelated form than not chelated (and animals can absorb them more readily through their intestinal walls).

Natural chelates are produced by soil microorganisms and are abundant in humus; they include various organic acids. The agricultural scientists, trying to improve on nature, have developed a host of synthetic chelates, the most commonly used being EDTA (ethylene diamine tetraacetic acid). These are sometimes promoted as "organic" chelates, which they are according to the chemical definition of organic—but they're not natural, and evidence is accumulating that they can be ineffective and have harmful side effects. Since the synthetic chelates are alien molecules, plants can only absorb them slowly. Also, after the chelating molecule releases its "payload," it may "latch onto" other nutrients in the plant, thus making them unavailable. For example, synthetic iron chelates cause a manganese deficiency and lower zinc and copper levels; EDTA "grabs" calcium ions and thus upsets the calcium-potassium balance. Synthetic chelates also put undesirable elements into the soil, namely sodium and chlorine, which are used in their manufacture. Finally, synthetic chelates are only stable under a very narrow pH range. All of which again only goes to show, "Nature's way is best."

In general, the plant nutrients are carried into roots along with the flow of water that we have seen earlier, independently of the laws of physics controlling water uptake, but controlled instead by laws governing colloids and ions. The special mycorrhizal fungi that were covered in Chapter 3 may help in absorbing phosphorus and nitrogen if they are present. Plant roots have an interesting layer of cells just outside of their "plumbing" region, called the endodermis, which acts something like a filter, regulating the passage of water and nutrients from the outside to the "plumbing" inside. The endodermis may help protect the plant against some toxic elements or help the plant to concentrate needed nutrients and reject abundant but unneeded ones.

Foliar uptake. Few people realize that plants not only take up water and nutrients through their roots, but also through their above-ground parts, especially their leaves. Plants can take minerals from the air in the form of dust particles and floating ions. G. Ingham of Pretoria, South Africa, believes the air contains all the nutrients needed by plants; a year's rain can deliver up to 30 pounds of nitrogen, 22 pounds of phosphorus, 338 pounds of calcium, and 41 pounds of sulfur per acre (*How to Grow Vegetables and Fruits by the Organic Method*, 1975, p. 107–109). These can be absorbed directly through the waxy outer cuticle which covers the outer cells or through the tiny openings on the surfaces of leaves, called stomata or stomates. Most plants have many tiny hairs on their leaves. It has been found by Dr. Philip S. Callahan (see his book, *Tuning in to Nature*, 1976) that plant hairs have an electrical charge because of their outer waxy covering, which acts as an electret (a material with a permanent electrical charge). He has found that similar wax-covered hairs of insects attract molecules out of the air. Could these plant hairs also act as tiny "antennas" to attract nutrient molecules from the air? Some research indicates that they do. It has been found that the hairs on tobacco leaves collect large quantities of radioactive lead and polonium particles from the air, increasing the carcinogenic potential of tobacco smoke (*Nature*, 1974, Vol. 249, p. 215–217).

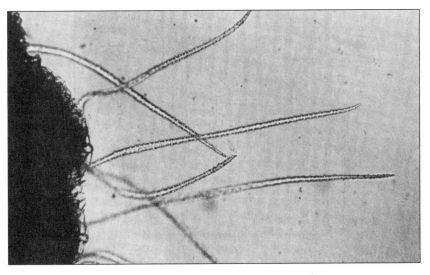

Alfalfa stem hairs (magnified 100 times under the microscope).

The ability of plants to absorb nutrients through their leaves is taken advantage of in the agricultural practice of foliar feeding, which can be a very useful and economical way to provide certain nutrients to crops, such as trace elements which may be deficient or to obtain a quicker response from the plant than can occur by root absorption.

Plumbing. But we have wandered away from our subject. We were being carried along with the water and nutrient molecules into the tubular "plumbing" (xylem tissue) in the root of our plant. The *xylem* tissue of plants is one-half of the plumbing system (or vascular tissue), the part which carries water and nutrients mainly upward in the plant. The other type of vascular tissue is *phloem*, which carries food and minerals from one place to another.

When water and nutrients get into the xylem, they can be quickly carried upward through the hollow, tubular vessels that form the "plumbing" by a sort of pulling force from above. Parts of the plant where water is needed (leaves and actively growing areas) create a water deficit, causing a pulling force on the water already inside the xylem vessels. So u-u-u-u-p we go!

Up the stem we rise. At every place on the stem where one or more leaves attach (called a node), there is a complex "traffic interchange," where the plumbing tubes of xylem and phloem divide into branches that go out into the leaves. As the growing season progresses, if a plant is not in good health, its "plumbing" can get plugged up in the nodes by mineral deposits, just as your house's plumbing can get clogged. As we will see later, this problem can be overcome by proper fertilization.

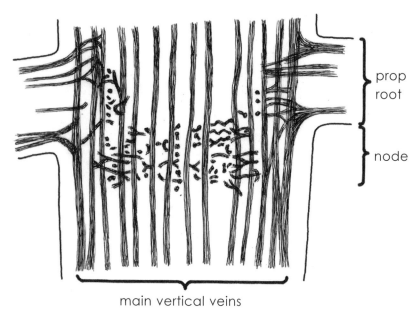

prop
root

node

main vertical veins

Thin vertical slice through a lower stem node of corn. Horizontal veins interconnect between vertical veins and between prop roots and other veins. Spots and dashes are veins that are turning corners and heading toward or away from the viewer.

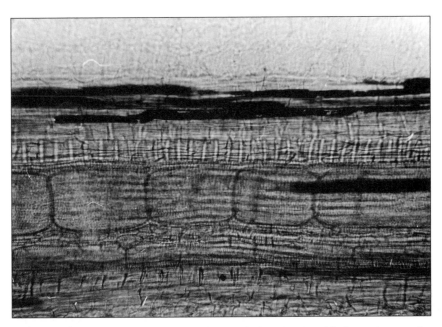

Xylem cells in a corn stem plugged with a dark deposit, magnified 100 times under the microscope.

The leaf. Up, up, up we rise until we go through one of the nodal interchanges and zip out into a leaf. As we work our way along, the veins of the plumbing get smaller and smaller until we flow out the very end of a xylem vessel and find ourselves in a new and wonderful world. It is a beautiful place, something like being in the downtown streets of a large city, because we are surrounded by row upon row of tall, crystal-like columns, mostly transparent, but studded with many emerald green oval objects, the chloroplasts.

We have finally arrived, like Dorothy arriving at the Wizard of Oz's palace, at the FOOD FACTORY, the awe-inspiring place where the food is made which ultimately feeds nearly all life on earth (except for a few types of bacteria which can live on inorganic mineral materials).

Within the chloroplast-containing cells of any green plant, from lowly algae to majestic redwoods, a miracle occurs—called *photosynthesis*. This is the dry, technical name for the plant's chemically combining of two common materials, water and carbon dioxide, into a life-sustaining molecule, sugar. In other words, the plant is a "chemical factory."

To understand more about this important process, we must digress to learn some more chemistry. There are different kinds of energy: light energy,

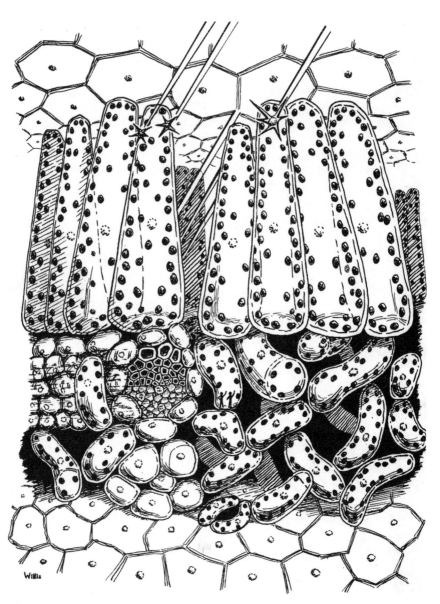

This view, drawn as if we were inside a leaf, shows the different parts. The upper epidermis like a transparent skylight lets light in. Beneath are the tall palisade cells where most food is made. The lower half of the leaf is mostly filled with loosely packed (to let air circulate) spongy parenchyma cells, where additional food is made. A vein (left center) supplies water and minerals and carries away food. At bottom center in the lower epidermis is an opening, the stoma, through which gases pass. Its size is regulated by two sausage-shaped guard cells.

heat energy, electrical energy, radioactive energy, and chemical energy. Our radios run on electrical energy, but our bodies "run" mainly on chemical energy—all living organisms and cells do. Chemical energy is the energy found in the various *chemical bonds* that hold together the atoms that make up a molecule or ion. When two or more atoms are held together, chemical energy is *stored* in their bonds, and when the bonds are broken, energy is *released*. To illustrate how much energy is stored in plant matter, just burn it; the chemical energy is then converted into heat energy.

Photosynthesis. In the "chemical factory" of a green plant cell, the raw materials of water (H_2O) and carbon dioxide (CO_2) are combined—this requires energy—to form the "manufactured product," a larger sugar molecule ($C_6H_{12}O_6$). The energy used to combine the water and carbon dioxide is light energy from the sun converted into chemical energy. Just as a factory assembles its product on an assembly line, the sugar molecule is not just put together in one step, but on a chemical "assembly line" in and near the green chloroplasts inside the cells. The chloroplasts are green because they contain the pigment chlorophyll. Chlorophyll is what "traps" (absorbs) the light energy from the sun. The sugar molecule (food) is thus a form of stored chemical energy. Incidentally, we run our automobiles on stored photosynthetic energy trapped thousands of years ago and changed into oil.

Some of the details of the assembly line of photosynthesis are shown in the accompanying diagram:

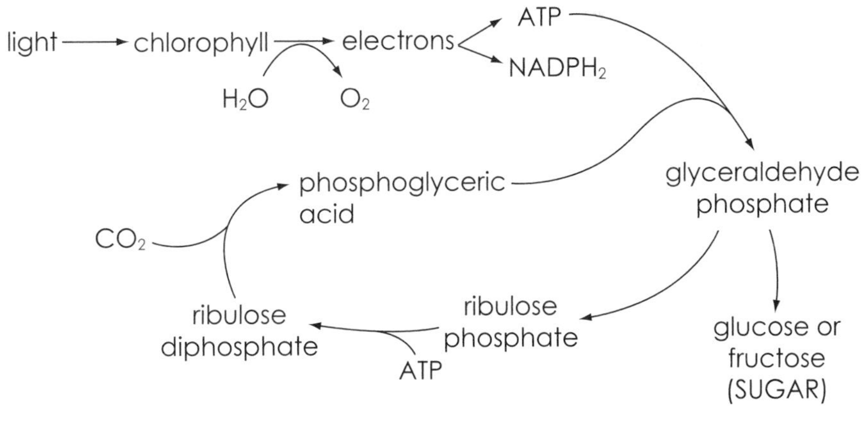

Four main chemical molecules are changed from one into the next in a cycle, with a molecule of sugar being "cranked out" at every revolution.

Two other components of the assembly line are vitally important: energy carriers and enzymes, for without these, the factory would grind to a stop.

Energy carriers. Chemical energy in cells is temporarily stored and carried from place to place in the cell by special energy carrier molecules. Some of the most common are ATP (adenosine triphosphate), $NADPH_2$ (nicotinamide adenine dinucleotide phosphate), and $FADH_2$ (flavine-adenine dinucleotide). These energy carriers contain one or more atoms of *phosphorus* as part of their molecule. AHA! Does that sound familiar—phosphorus? That's one of the macronutrients in plant nutrition! This is one of the main reasons plants (or animals) need phosphorus, to carry energy inside their cells. Phosphorus has been called the "energy currency" of plants (*Phosphorus for Agriculture*, Potash/Phosphate Institute, 1978).

Enzymes. Nearly all "assembly line" chemical changes in cells require enzymes, which are biological catalysts ("spark plugs"). Catalysts speed up the rate of chemical reactions. Otherwise the chemical reactions would not occur at a rate fast enough to sustain life. Enzymes are composed of a large protein molecule, plus usually another necessary part called a cofactor, coenzyme, prosthetic group, or metal activator. The cofactor can be either a metal ion (iron, copper, zinc, magnesium, potassium, or calcium) or an organic molecule (a vitamin or modified vitamin). Now do you see why some of the other plant nutrients, including some of the trace elements, and vitamins are so important in plant (and animal) nutrition?

There are thousands of enzymes in a single cell. Each kind of enzyme is very specific and can only affect one specific chemical change, one step of the assembly line. It is believed that an enzyme fits like a key in a lock on the surface of the substance or substances it controls (called the substrate). Actually, only certain parts of an enzyme, called active sites, are necessary for the enzyme to operate. The enzyme simply holds the reacting chemicals together long enough for the reaction to occur. An enzyme is not "used up" in the reaction; it is released to perform the process over and over. Enzymes can do a fantastic amount of work, from 10 to 10 million reactions *per second*!

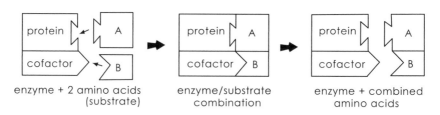

enzyme + 2 amino acids (substrate)   enzyme/substrate combination   enzyme + combined amino acids

Fertilize with CO_2?

Any required material which is in short supply or an improper condition that prevents a metabolic process from occurring as rapidly as it could is called a limiting factor. In the case of photosynthesis, inadequate light in the morning or evening or on a cloudy day would be a limiting factor, as would cold temperatures and lack of sufficient carbon dioxide (CO_2). The air only contains an average of 0.038% CO_2.

From an article by C.H. Wadleigh in the 1957 U.S.D.A. Yearbook of Agriculture, *Soil* (p. 41): "Under high light intensitites during the summer, this low level of CO_2 in the air [0.038%] probably is the main limiting factor in the photosynthesis. The production of an acre of a 100-bushel corn crop requires about 4 tons of CO_2. On a still summer day, the CO_2 level in the air of a cornfield drops very low. The corn under full sunlight could use 20 times the normal level of CO_2. We have evidence that the CO_2 produced by the respiration of microorganisms in the soil is an important factor in the supply of the gas to photosynthesizing plants: A well-fertilized soil rich in decomposing organic matter provides a much higher level of CO_2 in the air just above the soil than does a barren, infertile soil—proof that soil management can affect photosynthesis in the plant."

From the Oct. 1968 *World Farming* (p. 31): "Agronomists and even some farmers in the U.S. are making high maize yields even higher by adding carbon dioxide (CO_2) to their bag of practices. Carbon dioxide is a basic requirement for plant growth. Plants combine it with water to form sugars. A farmer in Indiana, U.S.A., tried reinforcing the air surrounding his maize field with carbon dioxide on the premise that on sunny days plants well supplied with water and fertilizer cannot always find as much CO_2 in the atmosphere as they could use. Adding CO_2 to greenhouse atmospheres to increase yields is already an accepted practice. But fertilizing with CO_2 in field crops is definitely exploring new territory. The Indiana farmer used dry ice as his source of CO_2, dropping one-pound cakes every 25 feet during the middle of the growing season. The small section of the field treated with dry ice yielded a startling 211 bushels per acre, compared to 158 bushels per acre on untreated maize. This doesn't prove CO_2 fertilization is practical, but the results are encouraging further experimentation."

Recent experiments have found that adding CO_2 to the atmosphere will not increase plant productivity as much as expected. Some crops could benefit, including soybeans, rice and wheat, but others, such as corn and sorghum, may not.

Respiration. The food produced by photosynthesis is used partly as a "fuel" to "run" cell activities, including some of the steps of photosynthesis. The stored chemical energy in food molecules is released in the energy-producing reactions of *cellular respiration*. During cellular respiration, the complex food molecules (carbohydrates, fats, and proteins) are gradually broken down in assembly line fashion (a series of chemical changes) to release the energy stored in their chemical bonds. The energy is then trapped and carried around by the energy carrying molecules mentioned before. The details of cellular respiration need not concern us here, but they are very complex, consisting of dozens of steps, nearly all of which require their own specific enzymes. How marvelously complicated is even one tiny cell!

Growth. The food produced by photosynthesis (glucose or fructose, called simple sugars) is used partly as "building blocks" to make new cells or cell parts, or to make special substances needed by the plant (such as enzymes, hormones, and pigments). Again, by a chemical factory assembly line series of changes, the simple sugars can be converted into more complex molecules, such as:

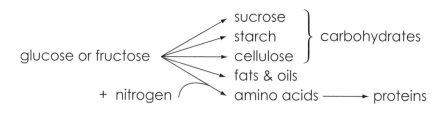

Again, specific enzymes are needed for the steps. Once more, do you see the importance of trace elements and other nutrients in the soil?

Translocation. The transfer of food and the other manufactured products just mentioned from one part of the plant to another is called translocation. It is very important to the living, growing plant because some parts of the plant produce or supply materials needed in another part. For example, food produced in the leaves goes to all other parts, especially developing seeds, fruits, or tubers; and hormones produced in one area affect processes in another.

Talking Trees, Sniffing Weeds?

We usually think of plants as rather inert organisms, just rooted in one place, quietly making food by photosynthesis. They don't seem to react quickly to environmental changes in the way that animals do, as when your dog or cat comes at your call, or runs away from a very large rottweiler.

A few exceptions have been known for a long time, such as the Venus fly trap, whose strange leaves close quickly to trap and digest insects, or the sensitive plant, *Mimosa*, a legume whose leaflets fold up when touched. Many plants produce special chemicals that evaporate into the air (called plant volatiles). The pleasant-smelling odors that flowers produce to attract pollinating insects are a well-known example. Strong-smelling chemicals found in leaves, such as in mints and citrus, act to prevent plant-eating animals (herbivores) from eating them. Other bad-tasting chemicals called phytoalexins are produced after an herbivore begins munching away.

In the last couple decades, scientists have discovered some remarkable new types of plant interaction. Some plants, including some forest trees, produce much larger amounts of phytoalexins in their leaves after an herbivore has eaten leaves of nearby plants. The "message" that warns the neighboring plants is through the release of one or more volatile chemicals which are carried on air currents. Tests found that the neighboring plants suffered less damage by herbivores and ultimately produced more seeds than unaffected plants.

Even more amazing is the recent discovery that the parasitic weed known as dodder (*Cuscuta*), which has no chlorophyll and must find a nearby green plant from which to "suck" nourishment, can detect odors of other plants and grow towards them. Tests at Pennsylvania State University found that up to 80% of dodder seedlings grew directly toward a nearby tomato plant, a suitable host species. But the seedlings avoided wheat, which releases a chemical that repels dodder.

Translocation over short distances can occur from one cell to its neighbor, but materials are mainly carried in that other half of the "plumbing," the *phloem* tissue. Just as the vessels of the xylem can become plugged up in an unhealthy plant, so also can phloem tissue, interfering with normal growth and seed or fruit production. Lack of available potassium causes corn nodes to become plugged by iron deposits; roots cannot receive adequate food and starve; yields are cut (E.J. Hewitt & T.A. Smith, *Plant Mineral Nutrition*, 1975, p. 171). So proper fertilization to maintain plant health is very important for producing high yields and high quality crops.

Hormones. One final digression before we complete our trip through the living plant. Most of us have a hazy realization that there are substances called *hormones* that regulate certain processes in animals and humans (such as growth and sexual functions), but few people know that plant growth and reproduction is also regulated by hormones, sometimes called growth substances. Hormones are chemical substances produced in small amounts in one part of the plant which move to other regions and act as chemical signals or messengers to regulate various processes. Known plant hormones include auxins, gibberellins, cytokinins, ethylene, and abscisic acid. They regulate such functions as cell division, stem-tip and root-tip growth and elongation, expansion of young leaves, fruit ripening, dropping of old leaves, and seed germination.

Sometimes man can use plant hormones or similar synthetic chemicals for certain purposes; for example, gibberellins may prove useful as growth stimulators in agriculture, while certain herbicides (2,4-D and 2,4,5-T) are synthetic auxins.

Reproduction. But back to our trip. We started out being carried from the roots up to a leaf, where we witnessed the marvel of photosynthesis

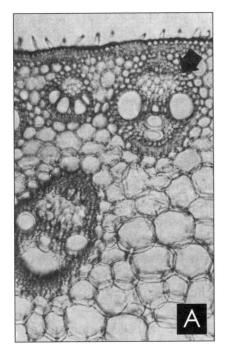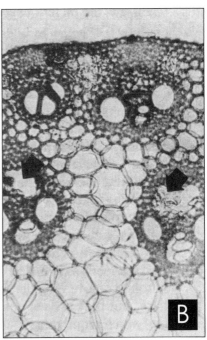

A. normal corn stem in cross section with several phloem cells (arrow) in each vein; B. corn stem with phloem disintegrating or missing (arrows) (magnified 100 times under the microscope.

and other chemical transformations in the "food factory." Finally, let's follow a food molecule into a developing seed or fruit, the reproductive parts of a plant.

In order for seeds or fruit to be produced, flowers must be produced, and in order for flowers to be produced, flower buds must first be formed. Tiny flower buds are formed long before they become obviously visible. In corn, the cob and tassel buds form when the plant is only about knee high. In apple trees, the buds that will produce next year's apples are formed this year. The number and quality of fruits or seeds depends partly on the number and health of the flower buds (plus other factors, such as weather, light, nutrients, and pollination). Some of these factors cannot be controlled by the farmer, but others can, especially the nutrition and health of the plant. Again, can you see why fertile soil is so vital?

After flowers have been pollinated, seed and fruit development begin. During this time, the nutrition and health of the plant are of utmost importance. There is a redistribution of water, sugars, amino

Soybean seedlings.

acids, organic acids, inorganic nutrients, and hormones from roots, stems, and leaves to the developing seeds and fruits. The need for some soil nutrients is at a peak at this time. Therefore the "plumbing" (both xylem and phloem) must be in good condition, as well as all metabolic and photosynthetic activities. In other words, the plant must be in excellent health. But as we have seen, both xylem and phloem can be plugged if the plant is not healthy. And some causes of poor health can be traced to the soil: insufficient nutrients, nutrients not available or not in correct form, improper balance of nutrients (various nutrients interact; too much of one compared to another can have adverse effects); and these situations can often be traced to insufficient humus and soil organisms.

So, from our place in the leaf, let's jump into a phloem cell and go gliding down the leaf, back into the stem, and out into a developing corn cob, for example. Here we are, and what a beehive of activity!

In earlier seed and fruit development, there is much cell division, producing new cells; later the existing cells simply enlarge as the seed or fruit swells. A great deal of food and other chemical substances from the leaves are moving into the seed or fruit for storage. Later, in the grain crops, much of the moisture in the kernels is (or should be) removed, producing a "dry" grain which can be stored without getting moldy. In both seeds and fruits, there are also chemical transformations in the type of stored food, from sugars to starches, organic acids, proteins, and fats or oils.

All of these metabolic activities require cell energy (supplied by phosphate-containing energy carriers), enzymes, and often hormones. And to occur normally and produce top-quality seeds or fruits, the plant must be healthy, with adequate water and nutrients. If there are trace element deficiencies for example, certain enzymes which regulate vital metabolic steps may be deficient, leading to low test weight or low bushels per acre, or low quality (low biological value) of animal feed. Or if the all-important "plumbing" is plugged, needed food may not enter, or water may not be removed to "dry down" grain naturally, requiring costly grain drying. These problems can largely be prevented if the soil is "healthy," with good tilth, adequate humus and soil organisms, and proper fertilization.

So here we are in a plant's seed or fruit. CHOMP! Oh, oh—look out, we're being eaten by an animal!

Chapter 5

Top of the Pyramid —

ANIMALS

T he food chain is often diagramed as a pyramid, called the biotic pyramid, with plants on the lower level, plant-eating animals (herbivores) next, and animal-eating animals (carnivores) next. If man is included, he is placed at the very top. The reason a pyramid design is used is to show that the numbers of species and individuals are greater at the lower levels and smaller at the higher levels of the food chain, because from one level to the next (when one organism eats or is fed by another), there is a certain amount of energy loss or wastage (food is stored energy).

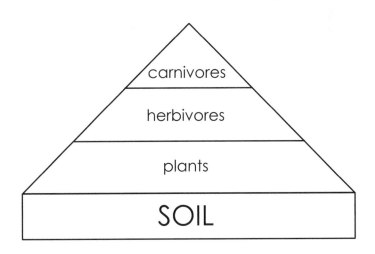

So when we move from the plant realm to the animal realm, we are moving to the top of the biotic pyramid. What affects animal (and human) health and nutrition? Are animals really that much different from plants? If you raise livestock or if you simply want yourself and your family to be healthy, you should know something about animal (and human) nutrition.

Sickness. Is it normal to be sick? Some people are ill so much of the time that they have forgotten what it was like to be well. Most farmers expect to have a certain number of sick animals and young ones that die. It doesn't pay to be sick. Doctors or veterinarians, drugs and medicines don't come cheap. In 2002, Americans spent $1.14 trillion for human "health care" (or is it really sickness care?), which was nearly 11% of the gross national product. Think how much money and time could be saved if you or your animals were never sick.

Unfortunately, medical and veterinary colleges spend relatively little time teaching future doctors how to keep their patients healthy, how to *prevent* disease. They almost entirely follow the logic of diagnosing a disease, giving it an unintelligible name, and prescribing some drug or poison to fight it, often by attacking the symptoms, not the real cause.

Sickness is NOT normal! Robust *health* is the normal condition of animal and human bodies. The dictionary defines disease as "disturbed or abnormal structure or physiological action in the living organism," and sickness as "any disordered and weakened state." We often hear the phrase, "you are what you eat." By now you know that animals get their food, which later becomes their body substances, from either eating plants or from eating animals that eat plants. And where do plants grow? In the soil. Dr. William A. Albrecht said it very well in his book, *Soil Fertility and Animal Health* (1958):

> We know that the soil grows grass; that the grass feeds our livestock; and that these animals, in turn as meats, are our choice protein foods. We can therefore, connect our soil with our health via nutrition. Since only the soil fertility, or that part of the soil made up of the elements essential for life, enters into the nutrition by which we are fed, we may well speak of animal health as premised on the soil fertility.
>
> There are increasing reports that animal afflictions are coming to be viewed as sins of omission, more than

that our livestock is falling prey to some stealthy force (possibly microbes). It is obvious that if an animal is deprived of some essential body constituent, troubles in its health will be encountered.

Animal nutrition. In some ways, animal nutrition is the opposite of the nutrition of plants which we saw in Chapter 4. Remember that plants take simple chemical substances (carbon dioxide, water, and minerals) and combine them into more complex substances (carbohydrates, fats, proteins), trapping and storing the sun's energy in the form of chemical energy. Animals, in their body and cell functions, first of all do the reverse: break down the complex food molecules, releasing some of the stored energy; but then they do as plants do, namely combine the simpler "building blocks" obtained from food substances into more complex substances that make up the cells and tissues of the animal's body. And you may recall that plants also can break down complex food molecules into building blocks and energy. Amazingly enough, many of the basic molecules involved in animals are nearly or exactly the same as those found in plants, such as amino acids, ATP, vitamins, etc., but many others are different, since plants do not have muscles, nerves, blood, etc.

So let's take a quick tour through an animal's body and follow a bite of food from a plant to see what happens to it. First an overview of what we will see: food is (1) digested in the animal's digestive tract, (2) absorbed into the circulatory system, (3) transported to all cells, (4) completely broken down to obtain energy or else assembled into new cell components, and (5) wastes are eliminated. Another complication in agriculture is that some animals have digestive systems pretty much like man's (pigs, horses, poultry), but others, called ruminants, have an unusual system with special features of the stomach (ruminants include cattle, sheep, and goats). We will first discuss non-ruminant (and human) nutrition.

Digestion. In most animals, food is first chewed in the mouth, which breaks it up into smaller particles to facilitate digestion and mixes it with saliva. Saliva not only makes the food easier to swallow, but in some animals also contains an enzyme that digests starch, called ptyalin. Upon being swallowed, we descend down the narrow esophagus and enter the large, sac-like stomach. Contrary to popular belief, less digestion takes place in the stomach than in the small intestine. In the stomach, the semi-liquid food-saliva mixture is churned and mixed with gastric juice, which contains strong hydrochloric acid and a few enzymes that act on proteins and fats. Preliminary digestion occurs, and the food is then pushed out into the long tubular small intestine, where most digestion occurs, plus absorption of digested food into the circulatory system. About 10 enzymes produced in the small intestine finish breaking down the food's carbohydrates, fats, and proteins. A summary of digestion is as follows:

starches
complex sugars $\longrightarrow$ simple sugars

fats $\longrightarrow$ fatty acids, glycerol, mono- & diglycerides

proteins $\longrightarrow$ amino acids

Undigestible materials move into the large intestine, where some water is absorbed to conserve body water. The large intestine also contains a multitude of bacteria, some of which are very beneficial in that they produce vitamins used by the animal's body, and in some cases they digest cellulose in similar fashion to ruminants' bacteria. Finally the waste matter, called feces, is eliminated.

Ruminants. Now let's consider those marvelous animals, the ruminants: cattle, sheep, and goats (plus members of the deer and camel families). These are the animals that "chew their cud." Their "stomach" is equipped with four compartments: the rumen ("paunch"), reticulum ("honeycomb"), omasum ("manyplies"), and abomasum. The first three are really expansions of the lower esophagus (called the forestomach), while the abomasum represents the true stomach, as compared to non-ruminant animals. A ruminant eats a large amount of plant food rapidly, swallowing it into the rumen. Then it is regurgitated and re-chewed ("chewing the cud"), and re-swallowed into the other stomach compartments. In the rumen and reticulum are billions and billions of special bacteria and yeasts, plus lesser numbers of protozoans. These microorganisms, especially the bacteria, are vital for the normal nutrition of a ruminant animal. First they digest the cellulose part of the plant-derived food (from the plant's cell walls), which most animals cannot digest, breaking it down and releasing the by-products of carbon dioxide, methane, and fatty acids, the latter being used by the ruminant as a source of nutrients. Also, the bacteria produce essential amino acids and many vitamins (those classified as water soluble vitamins) for the animal's use. They can also turn an inferior type of protein (lacking some amino acids) into a higher quality food source by changing one amino acid into another. These nutrients become available to the animal when the microorganisms are digested by the later parts of the digestive system. Thus, because of their microscopic helpers, the ruminants are virtually self-sufficient and can live on very poor quality feed. The use of antibiotics can kill or upset the balance of the microbes in the rumen; this can lead to stress, sickness, and even death.

Absorption. But back to the digested food molecules (simple sugars, amino acids, fatty acids, glycerol, and glycerides), plus the other needed food materials of water, vitamins, and minerals. In order to be used by the body, these substances must enter every living cell, which means they must be absorbed from the digestive system (stomach, small and large intestine) into the blood and lymph of the circulatory system, and be transported throughout the body, where they can enter whichever cells need them at the time. Any excess or unneeded food molecules can be either stored for future use or excreted as waste. The rate of absorption of one element is sometimes influenced by another; for example, a large amount of phosphate interferes with calcium absorption.

The efficiency of absorption by the small intestine is greatly increased by its millions of tiny fingerlike folds (villi) and their billions of microscopic microvilli, which add surface area to the intestine. It has been discovered that feeding animals harsh chemicals or drugs and antibiotics can destroy the microvilli and thus render the animal a "digestive cripple" for the rest of its life. This simply adds another stress to the body, which leads to general poor health or disease.

The products of carbohydrate and protein digestion (simple sugars and amino acids), and small fatty acids, are absorbed through the stomach or mostly the intestinal wall into the blood stream, but most of the products of fat digestion are first absorbed into the lymph and then flow into the blood.

Cellular metabolism. The blood carries needed materials—digested food, vitamins, minerals, water, oxygen, hormones—to or near to every living cell. They enter the cells and are used in the many marvelous activities of living cells. These can be boiled down into two categories: (1)

catabolism, or cellular respiration, in which the digested food molecules are further broken down to release the chemical energy stored in them, and (2) anabolism, or assimilation, in which digested food molecules are used as "building blocks" and reassembled into new cells or cell parts, or into useful secretions which some cells produce, such as enzymes, hormones, mucus, and milk.

All of the cell activities, both catabolism and anabolism, are together called metabolism, or metabolic activities. As we have seen for plants in Chapter 4, these metabolic activities, both breaking down and building up, proceed in assembly line fashion, in many steps, most of which require their own specific enzymes in order to occur at a sufficiently rapid rate. Again, the activities of metabolism are very complex, and we will not go into any detail here.

It is interesting to note that most of the steps of catabolism are identical in plant and animal cells. The energy released from food molecules can be used to "run" the various activities that any cell performs, including those of anabolism, or it may be used for the special job of certain cells, such as muscle and nerve activity. As in plants, the released energy is temporarily trapped and stored in energy carrying molecules, primarily ATP (remember, they contain phosphorus).

Hormones. In animals, the metabolic functions of cells are regulated by certain hormone-secreting glands: (1) the thyroid and parathyroid glands, which regulate general cell metabolism, plus the body's use of calcium and phosphorus; (2) the pancreas, which regulates the body's use and storage of sugar and starch (a malfunction of this gland can produce the well-known malady of diabetes); (3) the adrenal glands, which control the metabolism of minerals, water, carbohydrates, fats, and proteins, as well as preparing the body to exert extra energy in situations of fear, danger, etc.; (4) the pituitary and hypothalamus of the brain, which control the other hormone-producing glands; and (5) the gonads (ovaries or testes), which regulate reproductive activities as well as produce eggs or sperm.

Excretion. The final function involved in nutrition is the excretion, or elimination, of waste products. These include undigestible materials in the food eaten, digested materials that the body cannot use, and the by-products of metabolism, which are mainly nitrogen-containing urea (uric acid in birds) and carbon dioxide. These wastes are eliminated through four pathways: the intestine, the kidneys, the lungs, and in some animals, the skin.

Health. So we have completed our trip through an animal's body, following what happens to food molecules, or what should happen, for what we described is the *normal functioning* of the body and its cells. But just as plants can be unhealthy and have disturbed metabolism, partly from improper nutrients from the soil, so can animals also become sick, partly from improper nutrients in the food they eat. We humans certainly should realize that eating the wrong things or not getting enough vitamins and minerals will lead to health problems—the same thing is true of our livestock.

To be sure, other factors than feed can cause health problems in farm animals. One could say that sickness is caused by the body's succumbing to one or more stresses. Stresses can come from many sources: poor ventilation, impure or contaminated water, adverse weather conditions (temperature, humidity, precipitation, wind), infectious diseases, parasites, improper handling and injury, noise or other frightening conditions, lack of exercise, and of course improper nutrition. Most of these stresses can be reduced or eliminated by good farming practices and good animal management methods. Anything that we can do to reduce stresses will pay off in increased meat, milk, egg, or wool production, and reproduction of healthy young.

We cannot spend time on most aspects of animal management. What we are most interested in is good nutrition and how it relates to plant growth and the soil. For, just as is true for plants, an animal that is well nourished is better able to resist or tolerate such stresses as diseases, parasites, and adverse weather.

Nutrients. So let's back up and look in more detail at the nutrients needed by livestock and poultry. Remember, what are the two reasons animals need food—or to what two uses is digested food put in the cells? First, energy production, and second, building new cells, or growth and reproduction, plus repair of injury, and milk or egg production. Now let's look at the basic nutrients and see what their functions are. For energy production, the animal normally uses carbohydrates and fats as "fuels," although in starvation or excess, proteins can also be used. For growth, the necessities are proteins and minerals (calcium and phosphorus for bones and teeth, iron in the blood, sulfur in certain amino acids). But the cellular metabolic activities of both energy production (catabolism) and growth (anabolism) also require thousands of different enzymes, and as we covered in Chapter 4, enzymes are made of a protein plus a cofactor, which is either a mineral element or a vitamin. So here we see the need for vitamins and certain minerals, such as copper, zinc, iron, calcium, potassium, and magnesium. And then, certain minerals have other vital functions: phosphorus is part of the energy carrier molecules, sodium and potassium regulate internal osmotic pressure (water balance), chlorine and potassium regulate pH, iodine is part of the thyroid hormone, and so on. Finally, water is an essential nutrient substance since it is a necessary component of all living cells; in fact cells are mostly water. An average animal cell is about 70% water.

Requirements. Some nutrient substances are essential in an animal's diet, while others can be manufactured within the animal's body or cells. The following mineral elements are essential: calcium, phosphorus, sodium, potassium, chlorine, magnesium, iron, manganese, copper, zinc, fluorine, sulfur, iodine, and cobalt. About 10 or more of the amino acids are essential for most animals. Vitamins are essential. However, in ruminants and most plant-feeding animals, some amino acids and vitamins are provided by the bacteria living in the digestive tract. Most animals, other than man, monkeys, and guinea pigs, can manufacture their own vitamin C.

But it isn't as simple as being sure that an animal's diet contains the essential nutrients, for the nutrients must be present in the proper amounts and proportions to produce optimum good health. For example, a 1500 pound lactating cow may need 21 grams of calcium per day, but only 0.1 milligram of cobalt. Some elements are toxic in too large amounts, such as fluorine. There are interactions between some elements; for example in dairy cows, a calcium-phosphorus ratio greater than 2:1 can lead to milk fever.

Quality. So we see that for good animal health, not just any old food will do—we need *high quality food*. Food that will provide the right amount of the right kinds of nutrients in the right proportions. Are most livestock being fed really high quality food directly from the fields today? No, today we see the burgeoning practice of supplying feed supplements and trace mineral additives, because most farmers are not growing high quality crops. Their crops are incomplete, out of balance, and in some cases, even toxic.

How can we measure and evaluate the nutritional value of livestock feed? There are several methods.

Nutritional value. Three basic methods are used to determine *total energy* value of food. One is the total digestible nutrients, TDN. This is found by feeding a certain feed to an animal, collecting its feces, and finding the difference between the amount of nutrient fed and the amount in the feces. A "*digestion coefficient*" is calculated for major nutrients, including protein, nitrogen-free extract (sugars, starches, non-nitrogenous organic acids), fiber, and fat, as follows:

$$\text{digestion coefficient} = \frac{\text{intake} - \text{fecal excretion}}{\text{intake}} \times 100 = \% \text{ digestibility}$$

Then the percent of each nutrient in the feed is multiplied by its digestion coefficient and the figures for each nutrient are totaled to give an overall % TDN. The use of TDN to measure food quality has been around for a long time, but has some disadvantages. It does not work well for ruminants, and it does not measure all energy losses from the body, only the feces.

A better method of measuring food value is the caloric system. A calorie is a measure of heat energy; technically, it is the amount of heat required to raise the temperature of one gram of water one degree Celsius (1.8°F). The heat energy value of any food can be measured by placing a sample in a container called a calorimeter and actually burning it. The temperature rise of a water bath around the calorimeter can be easily measured and the calories calculated. Generally the number of calories in food is large, so the "large calorie," or kilocalorie (1000 calories) is used. It is sometimes written Calorie, with a capital C. The total energy given off when a food is burned in the calorimeter is called the gross energy. But this does not measure the amount of food energy the animal can actually use from the food. Again, a digestion trial is run and the animal's feces collected and analyzed. From this, *digestible energy* can be calculated: DE = gross energy – fecal energy. Digestible energy represents the part of the food energy which is absorbed from the digestive tract. But this is not the energy able to be used by the animal's cells, because some energy is also lost through the urine and from gases generated by rumen bacteria (mostly methane). So we can analyze the animal's urine and exhaled breath by confining it in a special cage, and then calculate *metabolizable energy*:

$$ME = DE - (\text{urine loss} + \text{gas loss})$$

The Calorie food values used in human nutrition are similar to metabolizable energy.

But metabolizable energy does not consider the amount of body heat lost by the animal. This can be measured by confining the animal in a large calorimeter. Finally, we can calculate the *net energy*:

$$NE = ME - heat\ loss$$

This method is the most precise way of determining food value. An article in *Hoard's Dairyman* ("Some Common Mistakes in Feeding Dairy Cows," Feb. 10, 1982) states that a high quality forage should contain 60 therms of net energy milk per 100 pounds of dry feed (1,000,000 calories = 1 therm). "Trying to get milk out of anything but high quality forages is futile and expensive."

A third method of determining food value is by a comparative slaughter-feeding trial. In this method, a number of animals are slaughtered before the trial to determine their body composition (either by chemical analysis or simply by finding the density of the carcass by weighing it in water). At the end of the feeding trial, the remaining animals are slaughtered and analyzed. The difference between the two groups of animals is called the *energy storage*. This method has the advantage of allowing the animals to be fed under more natural conditions than those in the calorie method.

Protein value. Another measure of food value is the *biological value*, BV. It is not a direct measure of the *total* food value, but a measure of protein quality. It compares the relative proportions of essential amino acids in the food with the animal's needs. A protein that lacks any essential amino acid is a poor quality protein. Since proteins contain nitrogen, BV is actually calculated as:

$$\frac{nitrogen\ retained}{nitrogen\ absorbed}$$

In other words it measures the nitrogen retained for the animal's growth, reproduction, or maintenance, which is determined by urinary and fecal analysis. Ruminants are special in that they have few amino acid requirements, since rumen bacteria can make amino acids from non-protein sources of nitrogen. Only cysteine and methionine are essential amino acids for ruminants.

Other measures of protein value of food are used, especially *crude protein* (CP) in agriculture. The name is deceiving because actually protein is not measured, but the food is "burned" to ash with strong acid and the

percent of total nitrogen found. This is then multiplied by 6.25 to give percent crude protein. This fairly simple method of finding food protein value is, as its name suggests, a *crude estimate*, since it does not consider the balance of amino acids nor the essential amino acids, and it is based on the *assumption* that the food nitrogen compounds are mostly the amino acids which make up proteins. But this may not always be the case, depending on the nutritional history of the food. In other words, there can be non-protein nitrogen (NPN) compounds in the food, such as amides, and a single nitrogen atom in the amino acid tryptophan, which non-ruminants cannot utilize to build proteins (except for horses, which have bacteria in their cecum [comparable to our appendix] similar to those of ruminants). Also, if plants are grown in soil having an excess of nitrates, as commonly happens when farmers apply too much inorganic salt fertilizers, the plant can contain a great deal of nitrates, which can even be toxic to animals.

Food quality. Thus we see that the *quality* of food, not just bulk, is extremely important in promoting animal health. And one of the most important aspects of food quality is the amounts of essential amino acids. Laboratory tests have shown that plants grown in mineral deficient soil

have an unbalanced amino acid production. In a test using peppermint, plants grown in calcium-deficient soil produced too much asparagine and proline, and too little glutamine, glutamic acid, and aspartic acid. Plants grown in sulfur-deficient soil produced too much asparagine and arginine, and too little glutamine, alanine, serine, glutamic acid, and aspartic acid (see box, Chapter 18).

But there is more than that. Nutrition tests of rats and chickens have shown that animals fed mixtures of pure, chemically synthesized amino acids do not grow as rapidly as those fed high quality natural proteins. So there are other nutritional substances present in natural foods which are needed for optimum health.

Even though ruminants and horses do not need such high quality protein as other animals do, the young of ruminants do need high quality protein to supply the essential amino acids because their rumens have not yet become activated.

Food efficiency. The "bottom line" of feeding agricultural animals is their efficiency in converting food into meat, milk, eggs, or wool. This goes beyond digestibility and assimilation into new body cells. The farmer is paid according to pounds and bushels. Quantity should not be the only determining factor in prices; quality should be considered too, but except for limited cases, that is how the market system works.

Different species of animals vary considerably in their food efficiency; for example a dairy cow, the most efficient farm animal, is 25% efficient in using food calories and 34% efficient in converting proteins into milk. A layer hen is 10% efficient for calories and 16% for proteins in egg production, while a beef steer is only 3% efficient for calories and 9% for proteins in producing meat. But there are other factors involved: the particular breed or variety of animal, presence or absence of the various stresses listed earlier (weather, disease, parasites, etc.), the genetic variation of different individuals ("ol' Bossie is just a better producer than Daisy"), and of course the *quality* of the food the animal eats.

A small increase in food quality, whether measured as digestibility or biological value, can cause a much greater increase in food efficiency and agricultural production, and in overall farm efficiency. For example, dairy cattle fed high quality forage give more milk, with a higher butterfat content, AND *eat less feed* while doing it! That's a situation most farmers would certainly like to be in, but sadly, most aren't. Why? It all goes back to the soil and proper fertility for the plants grown in the soil.

So, to complete our journey through the ecological food cycle, we make one more visit to the soil, to close the circle.

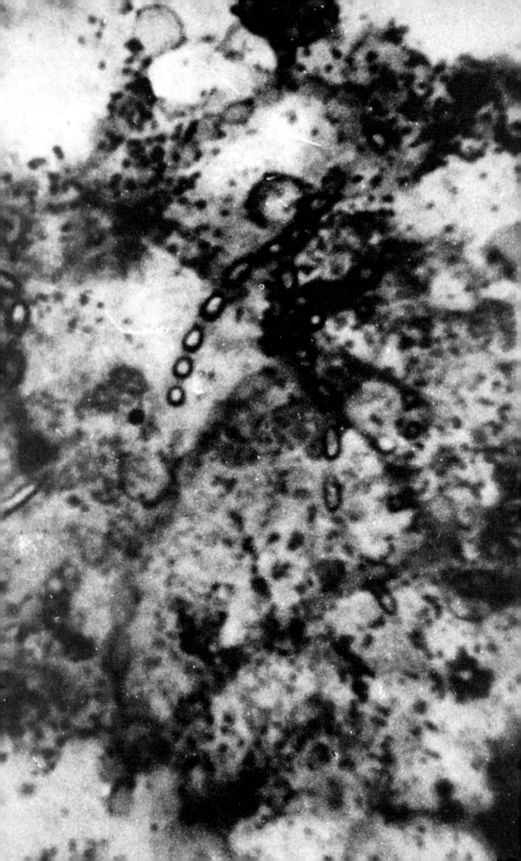

Chapter 6

Closing the Circle —

DECOMPOSERS

W ell, here we are down in the soil again. If it seems like we've been going in circles, you're right, we have! Remember, we are taking a trip through the food *cycle*. We now have to fill in the gaps and complete the circle.

In order to complete the food cycle, dead organic matter—dead bodies of plants and animals, plus the waste excretions of animals—have to be broken down and returned to the soil so that the nutrients they contain can be reused by plants—or to use the modern environmentalist's term, they have to be *recycled*.

There is a whole host of special organisms, the decomposers, who are ready, willing, and able to break down—to digest—organic matter. In doing so, they use it for food to build their own bodies or cells, but they later die or are eaten by other soil organisms, and eventually the nutrients are useable by plants.

We briefly met these denizens of the soil in Chapter 3. To review, they include both plants (bacteria, actinomycetes, fungi, and algae) and animals (protozoa, nematodes, earthworms, and arthropods, especially insects and mites). Some of them eat dead organic matter directly (bacteria, actinomycetes, fungi, earthworms, and some arthropods), while others are predators on them or each other (protozoa, nematodes, and most arthropods). Algae make their own food by photosynthesis like most plants. Some bacteria and fungi attack living organisms and may cause diseases and death of their hosts. A few bacteria can live entirely on inorganic nutrients, such as iron or sulfur compounds.

Soil Microbes and Crops

Over the last couple decades, microbiologists, including Dr. Elaine Ingham of Oregon State University, have discovered some interesting and useful details about soil microorganisms and how they relate to various types of plants growing in the soil.

Primarily focusing on soil bacteria and fungi, scientists have found that the more species present, the more productive the soil. Typical soil farmed by conventional methods may contain 5,000 species of bacteria, while good healthy soil has over 25,000. When soil is well-aerated and well-drained, beneficial bacteria and fungi will out-compete disease-causing species.

Ingham has found that the balance (ratio) of fungi and bacteria correlates well with the vegetation growing there. A weedy soil contains mostly bacteria, tall-grass prairie soil has about equal numbers of fungi and bacteria, a typical grassy pasture soil contains about twice as many fungi as bacteria, soil of row crops (corn, soybeans, sorghum) has about twice the fungi as bacteria, soil of perennial crops (such as berries) has slightly more fungi than bacteria, soil of deciduous trees (such as fruits) contains about ten times more fungi than bacteria, and a coniferous forest has a thousand times more fungi than bacteria in its soil.

Ingham has also found that application of salt fertilizers at over 100 lbs/acre at any one time kills some of the beneficial microbes, although several small split applications do little damage. Most pesticides kill some beneficial species and thus cripple their functions: recycling plant nutrients, controlling crop diseases, breakdown of toxic chemicals, and improving soil structure. A rapid (one year) way to restore crippled soil microorganism populations is to apply compost or compost tea.

So in the soil we have a whole "subcommunity" or "mini-ecosystem" of producers, consumers, and decomposers. We are mainly interested in the decomposers, and the most important are the microorganisms: bacteria, actinomycetes, and fungi. In addition, the earthworms do perform valuable services if they are present. They consume fresh organic matter and help "chew" it into finer particles which can later be more readily attacked by the microorganisms. Earthworms "haul" much surface organic matter down to deeper layers of the soil. They enrich the soil by their wastes and dead bodies, and they aerate and turn over the soil by their burrowing activities.

But now let's concentrate on the smaller decomposers, bacteria, actinomycetes, and fungi. We briefly described their characteristics and activities in Chapter 3. Bacteria are single-celled and can live in either aerobic (with oxygen) or anaerobic (without oxygen) conditions, but generally cannot live in very acid conditions. They help decompose organic matter to form humus, convert inorganic chemicals into useful plant nutrients (including nitrogen-fixation), and detoxify (break down) man-made toxic chemicals.

Actinomycetes are filamentous (thread-like) and require aerobic, neutral or slightly alkaline conditions. They aid in producing humus.

Fungi are filamentous, mostly aerobic, and can tolerate more acid conditions. Some help produce humus, but one type, the mycorrhizae, are symbiotic in plant roots, where they aid in nutrient absorption and root growth.

But now that that we've met the "work force," let's follow what happens when fresh organic matter decomposes. Most of the dead organic matter used in agriculture is of plant origin, either crop residues (dead stalks, leaves, roots, cobs, etc.) or animal manures (containing a large percent of partly digested plant matter). In recent years, some people have been using sewage sludge and papermill waste. So we are starting out with a food source for our hungry volunteer army containing a large amount of the components of plant cell walls: cellulose, hemicellulose, and lignin. These materials are very resistant to decay, particularly lignin, which can take months or years to break down. The rest of the plant material consists of whatever cell contents used to be in the plant, including sugars, starches, fats, and proteins, all of which are more easily digested. These initial components of organic matter have a high percent of carbon, compared to nitrogen, or a high carbon-to-nitrogen ratio, from 20:1 to 80:1 or sometimes more. Animal manures not only contain considerable undigested food, with the above constituents, but also millions of bacteria, both living and dead, from the animal's digestive tract, plus other miscellaneous substances, including vitamins and important mineral elements: calcium, boron, manganese, copper, and zinc.

Oxidation-mineralization. What exactly happens as dead organic matter decomposes depends on circumstances. If the material is left exposed to the air on the surface, where temperatures are not favorable for microbial growth and moisture is too little, a slow process of decomposition occurs called oxidation-mineralization. The elements of carbon, oxygen, hydrogen, and nitrogen, which make up carbohydrates, fats, and proteins, are released into the air as gases: carbon dioxide, water vapor, nitrogen, and nitrogen oxides. The inorganic mineral elements—sulfur, potassium, phosphorus, etc.—are returned to the soil. This process of oxidation-mineralization is very wasteful, for much potentially useful energy and nutrients are lost into the air; only the minerals are saved for later plant use. However, some advantages of leaving organic matter on the surface as a mulch do exist: protection from erosion, better water absorption and retention, moderating soil temperature, and weed control.

A much better use of nutrients can be made if the organic matter decomposes *in* the soil, not on it, for there the conditions of moisture and temperature are more favorable for the growth of microorganisms.

Microbial decomposition. There are two ways in which organic matter can decompose by the action of microorganisms: (1) anaerobic decomposition, or fermentation or putrefaction, and (2) aerobic decomposition. Anaerobic decomposition will occur wherever the organic matter does not contain much or any oxygen, such as inside a manure pile or deep in the soil (most soils are anaerobic below about 3–5 inches). As we shall see, anaerobic decomposition is also very wasteful of nutrients, or even toxic to plants.

Anaerobic decomposition. Typically, when organic matter decomposes anaerobically, it goes through two stages: (1) a brief period (about one week if moisture levels are adequate) called ammonification, during which certain bacteria release considerable nitrogen in the form of ammonia, NH_3. This causes the pH to rise into the very alkaline range (up to pH 8–9.5); (2) the second stage is called fermentation and lasts one or two months or more. More gases are released by bacteria and fungi: carbon dioxide, volatile organic acids, aldehydes, and alcohols come from carbohydrate breakdown; while methane, hydrogen sulfide, mercaptans, skatole, indole, amines, aldehydes, and others come from protein breakdown. Many of these have strong odors and cause the disagreeable odors associated with manure and putrid grain or silage. Some are toxic to plants and animals. The acids produced during this stage of decomposition lower the pH to 7–9. There is little cellulose or lignin breakdown in anaerobic decomposition (peat forms under anaerobic

Earthworms

Earthworms, often called "nature's plowmen," render a valuable service to man in helping to enrich soil. "Earthworm" is one of many common names used for various land-living segmented worms (Annelida); different varieties are often called angle worms, fish worms, night crawlers, and red worms. They are not at all related to the nematodes, or roundworms or threadworms, which are smaller and also found in the soil. There are giant earthworms in Australia, Africa, and South America which can get up to 7 to 9 feet long (think of the fish you could catch with one of those!).

Earthworms are scavengers, feeding mostly on dead or decaying organic matter, either by consuming it directly or by eating their way through the soil and extracting the small amount of organic matter in the soil. Their excretions are either plastered along their burrow walls or deposited on the surface as "castings." Analysis of earthworm castings reveals that they are richer in plant nutrients than the soil (about three times more calcium and several times more nitrogen, phosphorus, and potassium (K.P. Barley, *Advances in Agronomy*, Vol. 13, 1961). So along with the soil's army of microorganisms, earthworms help decompose organic matter and make the nutrients available to plants. Their "chewing" of fresh organic matter exposes more surface area for later microbial decomposition.

But that isn't all they do. Their burrowing activities increase soil porosity, aeration, and drainage, plus churning and mixing the soil. To see how they can mix soil, try placing several earthworms in a container that has the bottom half filled with sand and the top half with dark soil. In several months the sand and soil will be thoroughly mixed. Earthworms burrow several feet deep and can break up a hardpan layer. In one study, water equivalent to a four-inch rain was poured on two soils. Compacted soil without earthworms took over two hours to absorb the water, while soil with earthworms absorbed it in 15 seconds.

Tests have shown that crops grown in earthworm-inhabited soil increased yields from 25% to over 3 times more than in earthworm-free soil (K.P. Barley, *Advances in Agronomy*, Vol. 13, 1961).

Earthworms are sensitive to unfavorable soil conditions. They are rare in sandy soils and cannot live in strongly acid soils or in too dry or waterlogged soils. They are killed by excessive tillage, and by strong salt fertilizers and some pesticides; however, they can tolerate others, including DDT, but then birds are killed when they eat contaminated earthworms (*Pesticide Reviews*, Vol. 57, 1975). In general, earthworms can only do well in fertile soil, but proper use of inorganic fertilizers can build up soil humus (by producing abundant crop residues) and allow earthworms to become common. Dr. William Albrecht illustrated this by telling of a lawn around a new house that was seeded on "raw" clay with a thin layer of topsoil added. Ordinary N-P-K fertilizers and trace elements were added. Seven years later the lawn was dug up and there were numerous worm burrows, even down into the clay (*The Albrecht Papers*, 1975). Crop rotations that include forage legumes, grasses and cover crops encourage earthworms.

An abundant earthworm population is not only a sign of fertile soil—they are busily at work to make it even more fertile. If you don't believe it, prowl around your garden or lawn at night with a flashlight after a rain!

conditions). Anaerobic decomposition proceeds slowly and releases little heat compared to aerobic decomposition.

Aerobic decomposition. Under aerobic conditions, decomposition proceeds rapidly and releases considerable heat. Several types of microorganisms do the job, each taking a turn in breaking down the components of the organic matter. First, if temperatures are cool, certain fungi and bacteria called cryophilic ("cold-loving") bacteria (active at temperatures up to about 65°F), begin the work of breaking down carbohydrates (sugars, starches), fats, and proteins. The carbohydrates are mostly just "burned" for energy, with carbon dioxide and water being produced as by-products. Nitrogen from proteins and their amino acids is used to produce more bacterial cells. Heat is also produced as a result of metabolism.

Soon other bacteria and fungi appear, called mesophilic ("middle-loving") species, which live from about 40–110°F. They continue the decomposition and use the partly broken down intermediate products formed by the cryophilic bacteria as their food, along with the dead bodies of the preceding generations of bacteria. The populations of microorganisms increase at a tremendous rate, as does the temperature. Next, after several days, the thermophilic ("heat-loving") bacteria take over. They can only live at temperatures above 104°F. Their growth can raise the temperature in the center of a large mass of decomposing organic matter to 160°F or 170°F. After several days the temperature will drop and fungi and actinomycete species will break down some of the more resistant components, such as cellulose, hemicellulose, and lignin. The various microorganisms also produce enzymes which continue the decomposition even after they die.

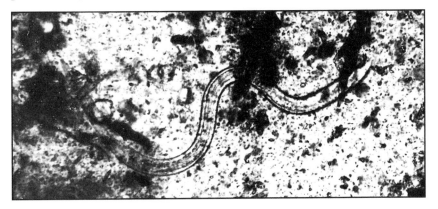

Nematode worm from decomposing manure.

During decomposition the ratio of carbon to nitrogen is reduced from its initial value of perhaps 30:1 or 50:1 to a final ratio of about 10:1 to 12:1. The reason is that the microorganisms use the high-carbon-containing compounds, carbohydrates and fats, for energy, and these materials disappear by being converted to gases: carbon dioxide and water vapor.

In actual field conditions, most organic matter decomposes under both anaerobic and aerobic conditions. On the outside of a mass of organic matter, or near a pore space in the soil, there will be plenty of oxygen, so aerobic organisms will flourish. Inside the organic matter or in poorly aerated soil, anaerobic conditions will exist. Actually, there are many bacteria which can function as either aerobes or anaerobes, called facultative anaerobes. In general, aerobic decomposition is to be preferred with regard to providing more nutrients to plants and not producing toxic by-products.

Humus. The end product of aerobic decomposition of organic matter is called humus. One ton of cornstalks, leaves, and cobs will produce about

100 pounds of humus. Humus is a complex, colloidal material (composed of small particles) containing proteins, lignin, fats, carbohydrates, and organic acids, as well as other broken-down products of the original organic matter, plus the metabolic products and remains of the microorganisms that did the decomposing.

As we have seen in Chapter 3, humus is a very valuable substance in soil in many ways:

1. It is a storehouse of plant nutrients. Because of its colloidal properties, it can hold three times the amount of nutrients that clay can.

2. It helps make nutrients available to plants. Acids released during decomposition (a) increase breakdown of mineral soil particles, (b) may form chelated compounds, which speed the plant's uptake of certain minerals, and (c) stimulate plant growth. Microorganisms temporarily "tie up" or hold nutrients in their living cells, thus counteracting leaching of the nutrients into the groundwater; the nutrients are later released for plant

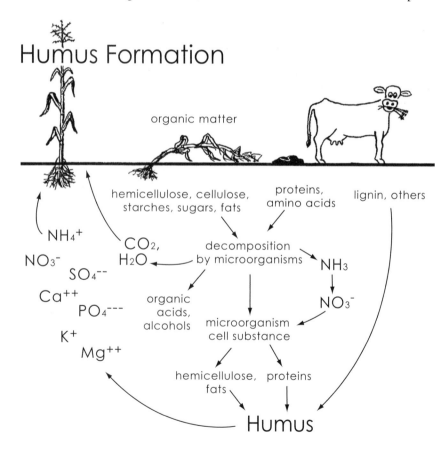

Humus Formation

organic matter

hemicellulose, cellulose, starches, sugars, fats

proteins, amino acids

lignin, others

NH_4^+

NO_3^-

CO_2, H_2O

SO_4^{--}

decomposition by microorganisms

NH_3

Ca^{++}

PO_4^{---}

organic acids, alcohols

microorganism cell substance

NO_3^-

K^+

Mg^{++}

hemicellulose, proteins fats

Humus

use when the microbes die. Some microorganisms, especially the symbiotic mycorrhizae, actually "feed" nutrients to plant roots, and of course nitrogen-fixing bacteria supply additional nitrogen. Certain fungi and bacteria protect plants from diseases and pests. The activity of soil microorganisms increases as the temperature rises; thus they provide more nutrients to plants at just the time the plants need them most: the warmest part of the growing season.

3. It improves soil tilth, making soil loose and crumbly. This contributes high water absorption and holding, good aeration, low water and wind erosion, and ease of plowing and cultivating.

Now let's apply these principles of decomposition to the practical operations of farming.

Composting. One direct application of microbial decomposition to agriculture is the practice of composting, which is especially used by followers of organic farming and gardening. Composting means letting organic matter rot *before* it is applied to the soil. The resulting product of decomposition is called compost, which is essentially, or nearly humus.

The proponents of composting are generally very enthusiastic about it, even though it requires extra time, equipment, and trouble to do. There are many methods of composting, which we do not need to consider now (see Chapter 13 for details).

Since compost is basically humus, it obviously is very beneficial to the soil. It is especially helpful to well-drained sandy soils, for it reduces drainage (leaching) more than adding fresh, undecomposed organic matter will. Compost has the advantage of being already prepared, "instant" humus. It provides its beneficial properties to the soil immediately, without having to wait several weeks or months for organic matter to decompose in the soil. And then, don't forget that not all soil is "healthy" (well aerated, free of toxins, full of beneficial organisms), so adding fresh organic matter to unhealthy soil will do no good and will probably do harm, because anaerobic decomposition will occur. Compost circumvents this problem, plus helps make the soil healthy and aerobic.

Another problem with adding fresh organic matter directly to the soil is that rapid microbial decomposition will "tie up" most of the nutrients (nitrogen, phosphate, etc.) before the crop plants can use them (as Dr. William A. Albrecht has often written, "The microbes eat first"). In other words, the microorganisms *compete* with the plants for nutrients when their numbers are large, but later when they die, the nutrients are made available to the plants. (This is less true of organic matter derived from legumes than from other sources because legumes contain considerable nitrogen.) For this reason, *large amounts* of fresh organic matter should not be incorporated in the soil just before or during crop growth. Fall application is preferred. But compost doesn't cause this problem and can be applied at any time.

Composting has another plus, if it is done at high temperatures (some composting methods use lower temperatures): it kills weed seeds and disease-causing organisms. But a minus is that some nutrients and carbon dioxide that could be used by crop plants are lost during the composting process. If the soil is "healthy" and fresh organic matter is incorporated in the right way and at the right time, results can be as good as with compost, or even better. So one should weigh the benefits of composting against the disadvantages in extra time and expense.

Animal manures. If done properly, the use of animal manures can be a very valuable asset to soil and crops. If used improperly, manure nutrients can be lost or soil and crops can even be harmed. Manure management will be covered in more detail in Chapter 13. In general, if manure rots aerobically it will help soil and crops by adding organic matter and nutrients. Tests have shown that mixing animal manures with inorganic fertilizers or rock fertilizers gives greatly improved yields.

Green manures & cover crops. The practice of plowing under a growing crop is in general very beneficial to the soil and future crops. Young growth is usually quickly decomposed, adding nutrients. Older plants are more resistant to decay and thus contribute more to the humus content of the soil. If a non-legume is plowed under, rapid decomposition can temporarily tie up nitrogen needed by crops, while legumes can increase the nitrogen content of the soil right away. Thus, a common practice is to plant a mixture of a legume and a grain or grass. Some species that make good green manure include alfalfa, alsike clover, barley, beggarweed, buckwheat, cowpea, domestic and Italian ryegrass, field brome grass, field peas, lespedeza, oats, red clover, rye, soybeans, Sudan grass, sweet clover, and vetches.

Planting such temporary crops in the fall (winter cover cropping) provides the additional benefit of protecting the soil from erosion over the winter. A possible disadvantage would be loss of more water from the soil by growing plants than from fallow soil.

Generally, green manures release their nutrients quickly and only benefit one year's crop, and contribute little humus to the soil, while animal manures decompose more slowly, providing nutrients for more than one year, and add more humus. An excellent practice is to incorporate a mixture of the two into the soil. One advantage of green manures is that fresh chlorophyll can absorb toxic substances in the soil. As mentioned earlier, organic matter should not be plowed in too deeply because below several inches, most soils are anaerobic and proper decomposition will not occur.

Detoxification. A few words about another valuable service provided by soil decomposers: the breakdown, or detoxification, of many of the toxic chemicals that man so liberally (and often unwisely) spreads around the environment.

We are all aware of the extensive use of herbicides and pesticides and that many of them are highly toxic, often to "good" animals and plants, even our crops (herbicide "carryover"). They also may pollute and accumulate in soil, surface waters, and groundwater. Fortunately, many herbicides and pesticides are not quite so dangerous and most do eventually break down, some more rapidly than others. Some become held (adsorbed) by particles of clay and organic matter, where their toxicity is greatly reduced. Some are broken down by sunlight or by chemical reactions in the soil. But a major method of breakdown is by the metabolic activities of living soil organisms. Soil bacteria are especially "talented" at degrading many pesticides, simply by using them as food. High levels of herbicides or pesticides will kill most or all soil life in the local area, but new organisms usually colonize the poisoned area from surrounding soil. Some toxic chemicals are readily broken down (such as the herbicide 2,4-D) while others are very resistant to microbial attack (the related herbicide 2,4,5-T). High soil humus levels (which favor microorganisms) are the most important property facilitating pesticide degradation. Unfortunately, modern farming practices generally reduce or destroy humus, as well as the beneficial soil organisms that produce it, while pouring on more and more poisons to kill the weeds and pests that result from destroying humus and soil organisms. Make sense? No, we're going around in a circle, but it's one of those bad kind—a vicious circle. The kind that close in on you and swallow you.

We have now come to the end of another circle, our journey through the food chain. It was a long trip, but an interesting and exciting one. One that revealed a great deal about how plants, animals, and the soil work. Also one that pointed out some problems. Now let's leave the world of nature and explore the realm of man's society. We may soon find ourselves going around in more circles!

Chapter 7

WHY PROBLEMS?

H ave you ever thought—on one of those days when both the tractor and manure spreader broke down, five cows are sick with milk fever, the land rental bill is overdue, and it hasn't rained for a month—that everything is against the farmer? . . . and that's one of your good days!

Surely the vast majority of farmers would not be doing what they are doing if they did not have a real love for the land, for raising crops and animals, for the (relatively) good clean lifestyle provided by rural living.

PROBLEMS! Everywhere you look, there are problems. The weeds and bugs are worse than they used to be. More cows (or pigs or chickens) are sick or dying than ever before. Expenses and taxes are higher than ever. Bank credit is tighter. More government regulations. Even the soil is getting harder and harder to plow. It's not just your imagination. Read some recent newpaper and magazine headlines: "Farmers Rally for Disaster Relief," "GM Corn Suspect in Illness," "Rainstorms Further Delay Harvest," "Help for When Your Income Drops," "Trade Deficit Hits Record," "Drought Help Available for Farmers," "Roundup Resistance Spreads," "Groundwater Declines As Much As 30 Feet Over Last Six Years," "Factory Farms Spread Bird Flu," "Iowa's Waterways Most Polluted."

Or consider these facts:

1. An average of 330 farms went out of business every week in 2002. Farmers comprised 30% of the U.S. population in 1920, 16.6% in 1950, about 3% in 1980, and about 1.7% today. A little over one-half of today's farmers are 45–64 years of age; only 6% are under 35.

2. Farmers' income derived from farming (not supplemented from outside employment) has fallen from 13.7% of the national income in 1947–1949 to 6.01% in 1980, and 2.75% in 2004, while production has steadily risen.

3. Of about 2 million farms in the U.S. today, 565,000 are family farms, but only 7% of family farms have totally on-farm incomes.

4. The farmer only receives about 10 cents of the consumer's food dollar. About 39 cents goes for off-farm labor, 8.5 cents for packaging, 4 cents for profits, 4 cents for advertising, 3.5 cents for energy, and the rest for several minor categories.

Even rural living is no longer what it used to be. Crime and drugs have invaded the country. Agricultural social workers cite farmstead health dangers of mental stress, lack of exercise, farm machinery accidents, toxic agricultural chemicals, cancer, and toxic gases in grain and liquid manure storage structures and livestock confinement buildings.

Why all these problems? And why are they getting worse and worse? We should first carefully analyze problems, note their symptoms, and determine their causes before attempting to solve them. And permanent solutions can only come from attacking the causes, not the symptoms; and once the causes are eliminated, the symptoms stop recurring.

Factors. In the complicated business of growing plants and animals, there are many factors and influences that can cause problems. There is the weather, which is pretty much uncontrollable by man (although there is growing evidence that man's activities do influence weather). Droughts, flooding, hail and windstorms can destroy crops. Then there are soil, plant, and animal nutrition. These factors are most certainly under the farmer's control. The genetic strains and breeding of crops and livestock can be chosen at will. Tillage and planting methods have a great deal to do with production, as do animal husbandry and other general management methods. The farmer can choose his time of planting, cultivation, harvesting, and his use of herbicides and pesticides, plus which animals to buy, sell, and breed. He can hire extra help, rent more land, build more barns and silos, and buy whatever machinery and equipment he wants (as long as he can get credit). He can usually market his products when and where he chooses. However, man-made laws and agencies do impose some restrictions on activities involving water, soil conservation, manure disposal, acres tilled, money available, and desirable times to market products. Certainly taxes impose a considerable burden. The overall economy dictates fuel, fertilizer, rent, and market prices.

In control? But to a very great extent, the farmer can be and should be in control of his situation and well being. But is he really? Or does he let others tell him what fertilizers to use and how much, which herbicides and insecticides to spray and when, where to sell his life-giving products, and what price he will receive? Advertising by business and industry plays a tremendous part in determining what brand of seed corn, herbicides, tractors, milking machines, and silos you buy. And even our neighbors influence us, for few farmers want to do anything much different from their neighbors although most do want to have a newer and bigger tractor. Unless your farm is paid for and you have been very lucky in weather and yields, your banker or loan officer probably plays a great role in literally running your farm. We tend to let the "experts" tell us how to till and fertilize our soil, treat our sick animals, and sell our produce. Is bigger always better? Are we devoutly bowing down in the temple of big agriculture to the high priests of technology, before the idols of 300 horsepower dual-wheel tractors and tall silos? Are we helplessly caught in the middle of a spider web of tangled threads? Besides the problems that unfavorable weather can create, are many if not most of our problems created, or worsened, by ourselves—by both farmers and those who advise or regulate them? Let's look at some possibilities.

Nature's problems. If you have read through Chapters 3–6, you should realize the vital importance of fertile "living" soil in producing high quality crops and healthy animals, for the soil is the absolute basis of agriculture. Anything the farmer might do that will harm his soil and soil life in any way will cause problems. And any other bad management practices involving crop and animal raising will add to those problems. Recognizing and solving these problems is what this book is mainly about, so more will be said in later chapters.

Man's problems. Would a university professor or extension agent or fertilizer dealer recommend a fertilizer program to you that will actually (1) kill beneficial soil organisms, (2) make your soil hard and more easily eroded, (3) destroy humus, (4) cause nitrogen to be released into the air and leach into the groundwater, (5) grow sick, poor quality crops, and (6) promote weeds, pests, and plant diseases? Impossible, you say? Would you believe they've been doing it for decades? As we shall see in Chapters 10–11, such commonly used fertilizer materials as anhydrous ammonia, muriate of potash (0-0-60 or 0-0-62), and dolomitic lime ("aglime"), plus over-use of many salt fertilizers, can do just that.

Cause & Effect

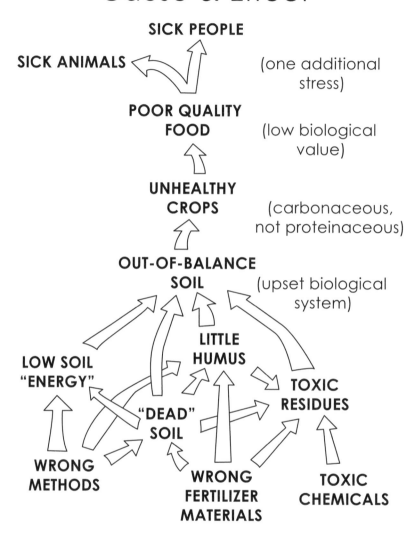

Can you believe that hybrid seed companies "design" their hybrid varieties for various types of "sick" and out-of-balance soil? Hybrid varieties are recommended not only on the basis of climate and desired maturity time, but also for high potassium or high magnesium soils.

Have you ever noticed how government programs and regulations so often seek to control us? Official "mission statement" descriptions of U.S. government agencies involved in agriculture often include such words as "protect," "promote," "support," "stabilize," "regulate," "manage," "facilitate," "supervise," "administer," "coordinate," "makes policy," "maintains orderly market conditions." And then, it seems as if the banking industry and the marketing system manipulate prices, interest rates, and credit to their own benefit. Is this just a persecuted figment of our imagination, or is there something wrong with our economic system? How many farmers have to buy the tractors and silos, or plant the number of acres that their bankers allow? Is this the land of the free, or not?

In a revealing study of the U.S. economic system in the mid-20th century (*The National Economy is out of Balance*, 1966), Dr. John Forbes proved by using government statistics that since about 1950, the economy has been growing increasingly unbalanced with a deceptive "prosperity" being fueled by unjustifiably high wage income and excessive private debt and interest. At the same time, farm income has gotten progressively smaller, proportionately after farm prices were taken off of 90% of parity and the "sliding scale parity" (90–60%) was put into effect. From the 1946–1950 period, when the economy was in balance, until 1964, the gross national product grew 149.5%, wages increased 164.9%, net private debt bulged 306.5%, net interest ballooned 523.3%, while farm income *decreased* 15.9%.

Dr. Forbes, a college history professor, was stimulated to make his study after hearing Carl H. Wilken speak out on the injustices to farmers. Wilken was an economic research analyst and auditor in the National Foundation for Economic Stability, Washington, D.C., as well as a farmer for 20 years. His story is told in the book, *Unforgiven,* by Charles Walters, Jr. Wilken's research into government economic statistics revealed what appears to be a constant relationship: that the nation's annual earned income is always about in the same ratio as the raw materials income (from agriculture [70%], petroleum and mining, fisheries, and lumbering). Thus, underpaying especially the farmer decreases the whole nation's income.

Both Dr. Forbes and Carl Wilken point out some of the man-made problems in our national economy and farm program:

1. Since 1950, farm income has not grown *in proportion* to the other sectors of the national economy.

2. This is because farmers have been underpaid for their commodities; that is, market prices have been too low.

3. Farmers are the only major producers in the nation who set no price on their products, but who accept whatever price the market gives them.

Farmers as a whole are unorganized and sell as individuals at wholesale, but have to buy supplies and equipment at retail.

4. Market prices have been deliberately held low to cause farmers to reduce production because of supposed "surpluses," which we are told are the cause of the problem. Government subsidies and production quotas are used to regulate production and marketing. Thus, the market is not free to operate on the law of supply and demand. Export of "surplus" farm products is being promoted to create a "favorable balance of payments" in international trade.

5. But "surpluses" do not really exist. Domestic production has not risen as fast as the population. Carl Wilken cited data from the 1944–1945 *Statistical Abstract of the United States* that in the 11 year period between 1933 and 1943, the U.S. imported more farm products than it exported, $12,786,725,000 worth of imports compared to $8,723,787,000 worth of exports.

The items imported do not consist only of products not grown in the U.S., such as coffee, tea, spices, and rubber, but imports of beef and dairy products have been steadily increasing, even though domestic production could easily supply the need. In 1962, 55% of agricultural imports were partly competitive with U.S. products. (*Farmer's World*, Yearbook of Agriculture 1964, p. 370). Compare these figures from the 1980 *Statistical Abstract of the United States*, p. 711:

Beef, in millions of pounds:	1960	1965	1970	1975	1979
Production	14,728	18,699	21,651	23,976	21,400
Exports	55	91	101	110	215
Imports	760	923	1,792	1,758	2,405

In the early 1980s only 1.2% of U.S. red meat produced was exported, rising steadily to 8.2% in 2000.

Part of the "surplus problem" is domestic under-consumption of certain items, related to the post-World War II shift in diet away from a wholesome grain-vegetable-fruit diet towards an over-processed "junk food" diet, high in sugar, fats, and meat.

In his comprehensive book, *Prosperity Unlimited* (1947), Carl Wilken stated that surpluses are needed for a prosperous economy, and they should not be sold at a lower price because they are surplus. "Without a surplus we would have nothing to trade for goods which we do not produce, nor would we have the materials with which to expand our economy so as to keep pace with population growth and technological improvement" (p. 8). "Scarcity of production is an economic vacuum which creates neither jobs nor income" (p. 102). "What to do with surplus production is a question that can be answered quite simply by the statement, 'Use it,' and thereby increase our standard of living" (p. 106). "Production creates its own demand through the income from production, processing and distribution" (p. 131).

Déjà Vu?

"Those who cannot remember the past are condemned to repeat it." So said the philosopher George Santayana.

It helps to be an older person, able to remember history that you have actually lived through. Nevertheless, when watching or reading the news today, do you have the uneasy feeling, "Hey, hasn't this all happened before?"

It really does seem to be true: history repeats itself. In fact, some believe that there are definite cycles of human activities, ranging from peace and prosperity to war and unrest, and connected to climate and astronomical cycles (see *Acres U.S A.*, August 2001, p. 28–29; September 2001, p. 30–31).

Consider the following near-repetitions of history:

1. The United States is sucked into a disastrous war in a third-world country, involving questionable politics, guerilla-style enemies and a polarized U.S. civilian population . . . Vietnam, and Iraq.

2. All-controlling corporate monopolies with vast influence over government, in an environment of little government regulation . . . the Robber Baron period of the 1890s, and the last couple decades.

3. Giddy speculation in stocks and real estate with loose government controls, unequal distribution of wealth with factory closings and layoffs, high national debt, interdependent world

economic structure, and low agricultural commodity prices . . . just before the 1929 crash, and today (2006–07).

4. And from an earlier time: military expansionism, a defensive stance against foreigners, economic decline from military expenses, high poverty rate, import of too many goods from Asia (loss of money from payments for imports), agricultural overproduction with farm failures, cheap food policy, easy government loans and subsidies, much money misspent on "pork" projects, government corruption, democracy lost to the common people, and senators giving up power to militaristic rulers . . . the Roman Empire before its fall, and the U.S. today.

Makes you think, doesn't it?

6. An economy financed by debt is not sound. Excess growth of credit and debt is the main cause of inflation. When borrowed money is spent in the economy, it does generate income, but does not produce new wealth, and the debt is still unpaid. New wealth and the new income it generates in the economy can only come from the production of raw materials. This would include the mining of ore and drilling for gas and oil, but those are non-renewable resources. Only raw material production involving the growth and reproduction of living organisms can produce an inexhaustible source of raw materials. This would include fishing and forestry, but especially agriculture. A farmer can put a grain of corn in the ground and in 80–120 days have at least one ear of corn with 600–800 kernels identical to the one planted.

7. The gross income earned by farmers for their raw commodities, when spent by them, generates about $7 of income in the economy for each dollar spent (this 7:1 ratio has held true since 1921; before that, one farm dollar generated less national income). This happens by a pyramid effect as the farmer buys equipment and supplies and stimulates the business of rural communities and the industries that make and distribute consumer goods, plus including the income generated as processors and middlemen market the farm commodities.

The income generated by the farm dollar is greater than that of any other segment of society. Compared to the 7:1 farm ratio, one dollar of wage income only generates about $1.56 of national income. For each dollar of new capital debt, about $4 of national income is generated, BUT it takes about $5.50 of national income to generate a dollar of profit with which to pay interest or liquidate capital debt. From 1929 through 1961, corporations increased their net worth 140.9% but their net debt increased 252%.

Our "Modern" System

Now, in the early 21st century, we are in difficult times. We are caught up in this "modern" western socioeconomic system. Agriculture is just one part of the system, and it is greatly affected by other parts.

The basic components of the system are (1) the *state* (government) and (2) the *economy* (capitalism). The state is dedicated to *control* of people (both its own citizens and even people in other countries) through laws, bureaucracy, the political system, police, courts, foreign policy and the military. The economy provides money to the state through raw material and commodity production and sales, services, banking and credit, investment, taxes and foreign trade.

Gradually building through the 20th century, the movement toward worldwide control of all sectors of society (globalism) has become a crescendo in the first decade of this century. The byword is expansion. Small corporations become big. Big corporations manipulate governments to give them subsidies, tax breaks and freedom from regulation, and to confer advantages over their competitors. They want everyone in the world to use their products and live their lives according to a uniform culture. A one-world, homogenized society is their goal. Critics and dissenters are put down.

In the food and agriculture sector of the economy, small farms are bought out (after following the chemical-intensive path) and conglomerated into vast tracts on which genetically identical monocultures are grown by caretaker "farmers." This "food" of doubtful quality is either force-fed to the inmates of livestock concentration camps or else is processed in a factory and transformed into something geometrically shaped called food.

Those who live in "developed" industrialized countries like to think they have a fair, democratic system, where pretty much anyone who strives can have some success and move upward. And compared to brutal totalitarian systems, our system is pretty good. When democratic (government from the people) systems are young, as just after a revolution, their idealism and newness usually produce a just, fair society. But then time passes . . . and corruption happens.

It seems like everywhere we look, our current modern system is greatly flawed: from crimes in high places (political, corporate) to cheaply-made shoddy products to poverty-level minimum wages to cheating home-repair contractors. Big corporations pay their top executives millions, hundreds of times more than the employees who really do the work, dump their toxic wastes into waterways if they can get away with it, and hire lawyers to avoid paying their fair share of taxes, and lobbyists to buy politicians. Politics and capitalism as they are practiced in the early 21st century are rotten!

How did we get here . . . and where are we going?

Is it really modern? The human species seems to be "hard-wired" to live in social groups, or tribes (to seek order and security), to interact and specialize (hunters, artisans, traders), to acquire property (or territory), to have leaders and followers, to establish rules of conduct and justice, and to get into trouble (greed and theft, cheating, murder and revenge). It's just human nature, we say. So, we naturally tend to form societies with imperfections.

Nothing new. Just about any trait of our "modern" system has existed in past cultures, from the most ancient on. "There is nothing new under the sun," the old proverb goes. The ancient Egyptians and Romans had political intrigue and assassination, cheating merchants and con-men, thieves and murderers. The Sumerians (about 2000 B.C.) had banking and accounting. The Persians (550–350 B.C.) had a uniform system of coinage. The Greeks (about 500–350 B.C.) had an affluent middle class and developed democracy, and the Byzantine Empire (about 500–

1453 A.D.) had a system of guilds (tradesmen and professionals) which set prices, wages, work hours and profits.

Modern capitalism (profit-seeking) emerged in Europe during the 17th century as international trading expanded after the discovery of the New World. The Bank of Amsterdam (1609) and Bank of England (1694) held public money and made loans (created credit) for commercial enterprises by issuing notes. Joint stock companies pooled the resources of many investors, and stock exchanges emerged. The growth of capitalism spawned government policies affecting money, trade, credit and capital accumulation.

The 18th century saw a dramatic growth in trade, colonial expansion, manufacturing and free-enterprise capitalism, including development of many advances in agricultural methods (plows, grain drill, crop rotation, selective breeding of livestock). Eventually these and other developments led to the Age of Enlightenment, new economic theory (Adam Smith) and the replacement of many absolutist monarchical governments with humanitarian democratic governments (the American and French Revolutions).

So, what we think of as our modern socioeconomic system has evolved gradually from very old roots. Many analysts believe that since the 1980s especially, our system has lurched alarmingly in the wrong direction. Free market capitalism as currently practiced appears to be driving us toward economic disaster, with conditions very much similar to those just before some past societies collapsed. Government deficit spending, personal debt, huge expenses for Social Security and Medicare, and pervasive waste in bureaucracy are danger signs. The extreme corruption of the U.S. political system adds to the danger.

Our modern system appears to be a fragile house of cards, and any of several unexpected factors could bring it down. Maybe Asian and other creditor countries will call in the billions of dollars in trade balance debt (loans) they have been buying. Will oil producing nations cut off our supply? Or maybe terrorists with nuclear weapons will launch a crippling attack. Perhaps a massive natural disaster will bring us to our knees (earthquakes, huge volcanic eruptions and tsunamis are quite possible). Remember the havoc caused by just one hurricane named Katrina.

Getting old? Many historians have observed that societies or civilizations go through a life cycle: birth, growth, peak, old age and death. Some think the United States and/or western culture has peaked and is in decline—economically, politically and morally. In his significant book, *Collapse: How Societies Choose to Fail or Succeed* (2004), Jared Diamond states that today's affluent society cannot last more than two more decades unless we make major changes, especially by solving the ecological problems of climate change, soil degradation, lack of clean water, deforestation and habitat destruction, resource consumption and waste production from increasing human population, invading foreign species, over-fishing, etc., as well as societal flaws of bad political leadership, failure to analyze problems and make rational decisions, and negative religious beliefs. He foresees either a catastrophic sudden collapse of industrial civilization or a future of significantly lower living standards. If the coming disaster is to be prevented, Diamond recommends long-term planning (unlike the typically short-sighted outlook today), learning from the mistakes of the past, willingness to alter our core values, and individually being a conscientious consumer and savvy voter.

Paul Krugman, in *The Great Unraveling: Losing Our Way in the New Century* (2003), thinks an economic bust will soon happen, with high interest rates and inflation, but eventually a moderate, good-willed majority will become upset enough to change the system.

Gar Alperovitz, in *America Beyond Capitalism* (2005), also predicts that our society will continue to deteriorate for about 10 years (income disparity, undue corporate influence over regulations, taxes, etc.), but when average people feel enough pain there will be a political and social revolution. The Democratic takeover of Congress in 2006 is probably not it; the corrupt system is still alive and well.

8. When farm commodity prices are held lower than a full parity level (in proportion with wages and retail prices of the rest of the economy), the national income is deprived of about seven times that amount, leading to a reduction in all the steps of the economy affected by agriculture. When farm prices were at or near parity level (before 1950), the economy was in balance, the national income grew but debt and inflation did not grow disproportionately. The nation was prosperous. When wages are allowed to grow disproportionately large and raw material prices are kept low, not enough national income is generated to pay for the increased wages, so borrowing must occur to make up the difference. This debt and its accompanying interest load, plus the reduced farm income, depress the economy, since consumer purchasing power is down. This leads to surpluses in the stores and on the farm, not surpluses from overproduction but from under-consumption. This leads to unemployment, more government money (from taxpayers and money borrowed against the future) spent on welfare and unemployment, and on farm price subsidies and storage facilities for the "surpluses." Taxes are raised to help pay for it all, and around we go! As Dr. Forbes put it: "The United States economy since 1950 has been the economy of a depression hid by debt."

9. Rather than blaming the farmer for "overproduction," we should look in the direction of the economists and advisers who set economic policy. We have been operating the economy under the theories sometimes called the New Economics, which are an outgrowth of the older Keynesian economics put forth by John Maynard Keynes in the 1920s. These economic theories state that capital should be devoted to (1) trade, (2) manufacturing, and (3) agriculture and raw materials production, in that order. Adequate farm prices are a hazard to earnings, but high wages are not. Cheap farm prices (cheap food) give people more income to spend on factory output. Dan Van Gorder in *Ill Fares the Land* (1966), summarized the Keynesian theory as "spend yourself out of debt." Keynes did not worry about inflation at all in his theory but thought that manipulation of tax and interest rates could keep inflation and unemployment in check.

Another aspect of the present economic and farm policy is the promotion of increased world trade (globalization), and "free trade," with a reduction or elimination of protective tariffs. That way our supposed agricultural "surpluses" can be shipped overseas to help keep the balance of payments more favorable (we have nevertheless been in the red in recent years), so we can buy more imports. That's fine if we import things we really need, such as oil (but do we really need so much of it?), but because our domestic food needs are not being met by restricted production quotas, we are also importing food unnecessarily. The recent international free trade agreements (NAFTA, CAFTA), promoted to lift up poor developing countries, have proven disastrous, bringing rich countries down as factory and professional jobs are outsourced.

Why are such illogical and obviously harmful and unworking economic policies being followed? Why are farmers most of all apparently being exploited and manipulated? In *Ill Fares the Land*, Dan Van Gorder states that the Keynesian economic theories promote socialism, and he traces them back to Fabian socialism of the late 1800s. This brand of socialism states that socialism will overtake the world gradually, rather than by revolution as in Marxism. Van Gorder believes that American agriculture was started on its downward path beginning with the New Deal programs and regulations of 1933. It was at this time that the "overproduction hoax" was originated to justify getting control over agriculture and to prevent farm bankruptcies during the Depression. Van Gorder even shows that the official government statistics were "revised" in 1933 to support the "overproduction hoax," and John Forbes also found that statistics going back to 1929 were changed without explanation in the 1966 "Economic Report of the President" compared to the 1965 "Economic Report of the President." Such tactics are apparently still in vogue; a syndicated newspaper article that appeared April 13, 1982, says that "the Reagan administration is using two sets of figures to estimate the 1983 budget deficit, depending on whether officials are talking privately to a roomful of powerful congressmen or to the general public." The George W. Bush administration has also been accused of the same thing.

Is 1984 Here Already?

George Orwell's chilling novel *1984* made waves when it was first published, and is once again being mentioned by analysts. It was not intended to predict what the world would be like in the exact year 1984, but Orwell simply reversed the last two digits of 1948, the year he wrote it.

We have all heard of some aspects of the book's plot: a world in which an authoritarian government tries to control all parts of people's lives, including speech and even thought control, through the constantly-repeated *big lie*, party *slogans* ("War is Peace," "Freedom is Slavery," "Ignorance is Strength"), *rewriting history*, and mass hysteria against a dangerous foreign *enemy*. The tactics of Thought Police and Big Brother is Watching You are official government programs, as is constant propaganda use of "Newspeak," the twisting of language to hide real intentions. A bleak world indeed!

Words, words. In his books, *Double Speak, New Double Speak* and *Double Speak Defined*, William Lutz chronicles the U.S. government's veering toward *1984* by use (and mis-use) of language. Since Richard Nixon's time, politicians and government officials have become masters of deception and rhetoric. Words with negative connotations are used for one's political enemies and euphemisms for one's own policies ("death tax" in place of estate tax, "revenue enhancement" for tax increase, "downsizing" for firing workers, "weak on defense," "cut-and-run," "flip-flopper"). Questionable political ideas are presented without nuance—it's either right or wrong, black or white, good or bad. Language and meanings are subverted and redefined, as in "we fight wars to bring peace, to make the world safe for democracy." And after a message is crafted, it is repeated over and over, especially in brief sound-bites and slogans, until it becomes "truth." They play on people's emotions rather than logic. By the way, such manipulation of language has long been used by public relations and marketing specialists, with interminable misleading advertisements.

Lutz believes we are trending toward a totalitarian system, and as Orwell portrayed, a totalitarian government is only possible when citizens' thoughts are controlled, which is much easier when people neglect critical thinking and the study of history, as is true today. Totalitarianism needs an object of hate, to keep the populace distracted from the failings and intentions of the government. Since World War II, we have lived pretty much in a permanent state of war.

Fascist? In an article titled "Fascism Anyone?" in *Free Inquiry* magazine, Lawrence W. Britt states that the U.S. is not yet a fascist state . . . but he cites such disconcerting historical signs of fascism as (1) strong nationalism and display of patriotism;

(2) obsession with national security; secrecy; (3) predominance of the military, with domestic needs ignored; (4) disdain for human rights; retaliation, imprisonment or torture of dissidents; (5) focusing on enemies as a scapegoat and diversion of public opinion from other problems; (6) controlled mass media, slanted to the right; censorship; (7) government promotion of capitalism and collusion with and protection of corporations, with labor unions suppressed; (8) interweaving of government and religion; (9) disdain for intellectuals, the arts and higher education; and (10) fraudulent elections, with media manipulation and altering election districts for political advantage.

Fascism is a right-wing phenomenon, first known by that name after World War I, when Benito Mussolini and his rightist groups, called *fasci,* took over the Italian government. The Nazi movement in Germany arose soon after. But the basic idea of ultranationalistic authoritarian government is not a 20th century phenomenon. Mussolini believed his movement was a legacy of the Roman Empire, and Hitler also believed that his Third Reich was a reincarnation of the Roman Empire. The military-led clique that took over in Japan in 1930 was not called fascist, but had similar objectives and tactics. They called their movement the New Order in Asia.

Communism, like fascism, is also an authoritarian and totalitarian system, but is left-wing in origin, and opposes true democracy, civil rights, capitalism and the rich class. It seems that human beings are capable of dreaming up a wide variety of social and governing systems, but going too far toward any extreme is most unwise.

Van Gorder says that the Keynesian philosophy is to socialize a country gradually, to allow private ownership but have government control. He cites general collectivist/socialist tactics of controlling the farmers first because being more independent, they are harder to socialize. He believes there is a concerted effort to force farmers off the farms and into the cities and factories, and to subsidize those remaining to produce less. Eventually, through higher prices, taxes, and government controls, free enterprise will be destroyed, and everyone will be made a ward of the federal government.

With the taking over of much of the U.S. federal government by right-leaning, ideologically-driven conservatives since the Reagan administration (some recent more liberal control has only temporarily interrupted the march of conservatism), various analysts and authors have stated that there is a definite "program" or agenda of returning American society to early 20th century pre-New Deal days, with a greatly limited government, low taxes for corporations and high-income-bracket persons, and gutted social and environmental programs.

There is a group of 20,000–30,000 right-wing ideologue lawyers and judges who in 1982 founded the Federalist Society and who have been influencing conservative politicians to accomplish these goals. So-called neoconservatives clearly said so in a 1997 document called "Project for the New American Century," and other statements on their website, www.newamericancentury.org. Their principles include greater American militarization, challenging hostile governments, and advancing democratic and economic freedom. They say, "American leadership is good both for America and for the world." Their tactics appear to include dividing the electorate with political wedge issues, increasing the growing rich-poor wealth gap, purposely bankrupting the country with deficit spending so social programs can be cut, and controlling the people through taking away civil rights and by fostering fear and insecurity. Insiders are put in office sometimes by subverting fair elections, and cronies are appointed to the bureaucracy.

The result? Politics is defiled by special-interest money. Corruption of every sort is rampant. Some observers believe that international bankers and corporations, along with a few rich dynastic families and influential intellectuals are seeking to basically control the world, perhaps with the good intentions of bringing peace and prosperity to all, but with very questionable effects. See such books as *Confessions of an Economic Hit Man* (2004), *America Beyond Capitalism* (2005), *Blowback: the Costs and Consequences of American Empire* (2000), *The Great Unraveling: Losing Our Way in the New Century* (2003), *Nemesis: the Last Days of the American Republic* (2007), and *Dangerous Nation* (2006).

Whether or not you want to believe in the existence of a worldwide or a national conspiracy, you must admit that many organizations and policies in the last several decades have brought more and more regulation and control upon us, and the farmer hasn't exactly fared well.

In-house Criticism

Besides the growing number of socially and environmentally oriented critics of our modern, profit-minded big agricultural system, a few academics inside the universities are also speaking out on the insane path we have been going down.

In a 1998 article in the respected *Agronomy Journal* (Vol. 90, p. 1–2), Iowa State University agronomy department professor E. Charles Brummer said that although the American agricultural system *seems* to be highly successful and productive, "closer inspection reveals a highly unstable system, easily disrupted by any number of factors, including the weather, insect and disease epidemics, and foreign trade vacillations." He notes that since the 1950s, emphasis has been on making a few crops as productive as possible, grown in monoculture, rather than building a stable system.

Brummer lists several failings of conventional agriculture: (1) in spite of reduced erosion from no-till methods, still we are losing soil faster than it can be regenerated; (2) there is still considerable pollution of land and water from excess fertilizer and toxic chemicals; (3) concentrated livestock feedlots do not allow manure to be easily recycled; (4) fifty-plus years of chemical weed and pest control have only made weeds and pests harder to control; and (5) although gross farm receipts have been rising, higher input costs have reduced net income.

Brummer believes that these problems can be overcome and that a stable system can be built by increasing *diversity,* including raising a variety of crop and livestock species, crop rotation, multicropping and intercropping. He also notes that organic methods have been proven feasible even on a large scale.

Next witness. In 2004, Charles A. Francis of the Agronomy and Horticulture Department, University of Nebraska, wrote a scathing analysis of the entire economic/corporate/agricultural food system *(Agronomy Journal,* Vol. 96, p. 1211–1215). He begins, "We live in a U.S. culture that is driven by short-term economics ... in contrast, the time frame for sustainability is forever." He then notes that national economies are becoming one global system, with multinational corporations accountable to their stockholders, rather than to the general public. Because of increasing awareness and criticism by the public of unequal wealth distribution and environmental damage, agricultural corporations are responding by polishing their image and "greening" some of their operations. But is this a real conversion or merely a "greenwashing" of their image?

Francis covers five negative trends in the global food system:

1. Consolidation of former small family farms into huge operations that plant monocultures and use animal confinement units, with resulting loss of diversity, jobs and rural communities. Agricultural economic studies have found that for field crop production, the most efficient farm size is about one section (640 acres), while most of the profits from large-scale operations bypass the farmer and enrich the corporations that sell the expensive inputs. Family farms have a long-term focus based on stewardship, and their income then cycles several times through the local community.

2. Genetically modified organisms may have some short-term advantages, but weeds and pests can develop resistance, and the foreign genes can escape to wild species and produce super-weeds or pests. Corporate ownership and monopoly of genetically engineered plants and animals, and the high costs involved are also worrisome.

3. The buying up of small organic companies by big multinational corporations leads to abandonment of regional food production, loss of local jobs, and possibly loosening of organic certification procedures.

4. An increasingly globalized food system is leading to a homogeneity of foods and diets, loss of local identity, dependence on long and fragile supply lines, and exploitation of workers in developing countries.

5. The global adoption of capitalism assumes that the free market system really works. But unfair competition and dumping of commodities below the cost of production have richly rewarded corporations and impoverished local farmers in developing nations—a form of global economic colonialism. The current

economic system is focused on accumulating individual wealth by landowners and financiers. Money is the only medium of investment and return, and environmental costs and damage are ignored and passed on to future generations.

Francis recommends "a thoughtful sorting out of our needs versus our wants," a system that considers limited natural resources and their sustainable use, support of local economies, and a modest amount of trade.

Still more. In a 2005 article in *Science* (Vol. 310, p. 1621–1622), Julie Wrigley of Stanford University and seven other academics from U.S., Canadian and Italian institutions, lay out a clear indictment of the "factory" system of livestock production, with large-scale confinement raising of cattle, hogs and poultry. They state that such methods produce three-fourths of the world's supply of poultry, over two-thirds of eggs, and 40% of pork. They note the increasing trend away from livestock grazing and toward feeding of grains, with the former close link between raising meat animals on the land, close to the consumer, being broken. Factory-style livestock operations require large amounts of water, especially for feed production, and then manure accumulation contaminates water by releasing nitrogen, phosphorus, antibiotics and disease pathogens. The poisoned water can end up in the ocean and cause "dead zones." There is no economic payment for the environmental costs of confinement operations.

The authors say that these negative effects need to be fixed. There could be surcharges to pay for manure disposal, and tougher water quality regulations. Stronger consumer demand for ecologically-sound meat would have a great effect. Finally, they state that a "strong political will" is needed.

Solutions. These man-created problems are so great and daily getting worse, that one might like to throw up one's hands, say "What"s the use?" and give up farming. Many have already. Don't give up hope. There are solutions.

In their books, mentioned earlier, Dr. Forbes, Carl Wilken, Charles Walters, Jr., and Dan Van Gorder give some steps which could go a long way toward correcting the problems:

1. Restore farm prices to a parity level and bring the economy into balance, with the different sectors growing proportionately. Eliminate subsidies. A full parity program will not cost money since farm income generates about $7 for each $1 spent. Parity prices don't lead consumer prices, they follow them.

2. Establish a farmer-controlled food processing and distribution system.

3. Develop local markets in cooperation with business and civic or organizations.

4. Maintain protective tariffs at the parity level, representing the difference between the American price level and the price of imported products at the point of entry. When the world price reaches our level, all tariffs could then be at zero, or on a free trade basis.

5. Diversify crops in an area; suit them to soil and climate. We could even grow coffee, tea, spices, and rubber in our southern states and become nearly self-sufficient.

6. Use farmers as policy advisers; have farm referenda before making laws affecting farmers.

7. Strengthen family and youth activities to keep youth on the farm.

8. Return 15–18 million urban people to the land (those with previous farm backgrounds).

9. Break away from reliance on herbicides, pesticides, and commercial salt fertilizers and re-establish ecologically sound agriculture.

10. Educate people to the fact that farming is our most important industry.

A real solution to the farm problem, and many others, cannot come until those who are in charge, who are making policy and directing programs, and who are working in manufacturing and business, are not motivated by such human failings as greed and lust for power and self-importance.

Your part. We have seen some disturbing facts and ideas. You may feel pretty helpless to change the national farm policy and economic system right now today—so what CAN you do to help solve problems? As an individual farmer, you can certainly change those things you do have control over: your own farming methods—your fertilizing and tillage methods, crops grown, animal management, machinery owned, and even marketing. The remaining chapters will show you how.

Starting to Learn,
but for the Right Reason?

For the last few decades of the 20th century, the "agricultural establishment" (government agencies, university professors and researchers, big ag corporations, local ag supply companies and their field agents) was either ignorant of sustainable/organic methods, or else decidedly hostile toward them. The establishment did no serious research on natural methods or inputs, kept pushing industrial-scale operations and chemical-intensive methods, ridiculed the practicality of alternative methods, and worked openly or behind the scenes to undermine or destroy any competition. U.S. Secretary of Agriculture in the early 1970s, Earl Butz, said, "Get big or get out."

Such tactics continue today to some extent, but a combination of all sorts of disasters caused by conventional methods, plus growing awareness and popularity of natural, clean food have made the "poison-pushers" give natural methods some grudging respect. Many colleges and universities have professors and courses dedicated to sustainable agriculture. Federal and university researchers are investigating such subjects as earthworms, beneficial soil microbes that attack weed seeds or protect crops from diseases and pests (replacing fungicides and insecticides), use of natural volatile plant compounds to fumigate soil instead of very toxic methyl bromide, humus and compost, use of cover crops to replace some crop nutrients and improve soil structure, reducing nitrogen fertilizer through split applications, and comparisons of sustainable and conventional systems. In many cases, they are merely rediscovering what natural-system farmers have known for years—the scientists may only be adding a few decimal places to common sense "farmer knowledge."

And then the big ag corporations are using whatever they can from this research to develop patentable products they can sell to gullible people, such as genetically-engineered nitrogen-fixing seed inoculants, and herbicides, fungicides or insecticides (biopesticides) using extracts of microbial or plant-produced chemicals.

And in time-honored big-business fashion, the large companies often try to buy out small alternative companies, not always to adopt their products, but to render them extinct. They are desperate to paint themselves green. Large confinement animal operations try to use "organic," "pastured," or "grass-fed," even though they bend the requirements for those categories near to breaking. Big-box stores are introducing organic products, while the corner grocer has been stocking them for years.

Chapter 8

CHANGING COURSE

A re you fed up with things the way they are? With decreasing crop quality, sick animals, weeds, bugs, higher fertilizer and veterinary bills and lower market prices, and tighter bank credit? Yes, but are you REALLY fed up? Enough to actually DO something about it?

You don't have to be a slave to the conventional system of agriculture, with its over-emphasis on technology and quantity, and its neglect of natural systems and quality. You *can* break away. You *can* change.

Wouldn't it be great to have loose, crumbly soil that didn't wash away in an average thunderstorm? To stop having to poison your land (and yourself) with insecticides and herbicides? To need less fertilizer each year, not more? To grow such high quality crops that you can actually feed less feed to your animals and get more production, and have few if any sick animals? To be able to rent less land and grow more? To be able to market your produce where you want, when you want, and at your prices? It's not a dream. *Farmers are doing it now!* And they don't necessarily have rich, loamy soil to start with, either—sometimes just "blowsand."

If you really do want to change your farming methods—or even just seriously investigate the possibilities—what do you have to do?

Is it right? The first thing is to become fully convinced in your mind that natural systems farming is the right way to go. Does it make any sense to use dangerously toxic insecticides on crops that people or animals are going to eat? Do you like being locked into a certain crop rotation because herbicide carryover is so great? Does it make sense to use fertilizers that sterilize the soil of its beneficial organisms and destroy vital humus? To grow crops that are so sick and lacking in nutrients that not only do they succumb to pests and diseases, but they contribute to the ill health of animals and humans?

Does it pay? Isn't it more desirable to nurture and care for what provides you and ultimately all Americans their living—your soil? Yes, you say, but does it pay? Can I make it financially by farming the natural-systems way? Absolutely. Farmers are doing it now. Who can argue with higher yields and better quality (premium market prices) plus lower herbicide and pesticide, veterinary and doctor bills? It is true that during the transition period when you change over from your present methods to natural system methods, you will have to delicately balance purchases of new fertilizers and equipment (if needed) against lowered chemical and medical bills. But encouragingly positive results will be seen right away the first year.

Scientific studies comparing conventional farming methods with "organic" and Amish farming methods (comparing mainly dollars, energy, and yields) have found that the more natural methods are definitely economically feasible, especially with higher fuel, chemical, and fertilizer costs. Some specific examples are given in Chapter 15.

Learn. After you have decided that natural systems farming is the way to go—or at least worth trying—what next? Next you should learn as much about the subject as possible, so that you not only will know what you are doing, but why. It is hoped that this book will be a good start in that direction. Other helpful resources can be found in Appendix B.

Seek advice. Unless you are very knowledgeable in all aspects of soil chemistry, plant physiology, and animal nutrition and health, and can keep up with the latest developments and research, you should seek the advice of those who are. But you have to be careful. Can you trust the advice they give? Some advice-givers are more interested in selling their product than in really helping the farmer. Others are simply misinformed or caught up in the high technology/low quality "science-will-save-us" syndrome.

Listen carefully to what they say and compare it with what you know about natural systems farming. Do they recommend toxic chemicals to "solve" weed and pest problems, or do they try to get at the real cause of the problem? Do they recommend fertilizers that harm the soil and soil life (see Chapters 10–11)? Do they try to build up the natural systems of the soil-plant-animal-soil cycle? Are they concerned about quality, not just quantity?

Check with their customers and see what results they have gotten. Are they honest in their promises? If they guarantee a certain yield figure, watch out, because no one can guarantee a yield without being able to control the weather.

Costs. Don't be afraid of a reasonable fee for consulting services. After all, you are paying for their "expertise" and advice, based on considerable experience and knowledge, and if it gives you just a few more bushels per acre, it pays for itself. And that doesn't include the value of improved soil, higher quality crops, and healthier animals.

It's entirely possible to go on a natural systems program and spend no more than you had been previously spending for fertilizers and chemicals; and less if you figure in veterinary and grain drying bills.

A step at a time. The steps which should be followed to turn your soil around and begin restoring natural systems are simple. First, your soil should be tested to see where it is now, what its levels of nutrients and other characteristics are. This is like when you have a fever, you put a thermometer in your mouth to see how bad it is. Only when you know what's wrong can you start to identify problems, find causes, and find solutions.

Next, you (or your consultant) should begin correcting the problems by attacking them from two directions: (1) stop doing wrong things, and (2) start doing right things.

Of course, it's not really that simple, because you have to know what are the right and wrong things, and how and when to use the right ones. This is where the advice of a good consultant is invaluable. As we shall see in later chapters, various fertilizers and soil conditioners can have widely different effects on soil and crops, and even the good ones can be misused. And then they can interact—one affecting the other—if used together.

Also, don't forget that your soil's problems may have taken decades to develop, so you can't expect to solve them overnight. Depending on your soil type, the weather, and your budget, it may take several years to make real progress. But as we have said, positive results will occur the first year. And every year after that will show improvement.

So that's what you need to do to start changing course. The only thing holding you back is you—so let's get up and go!

Chapter 9

TESTING, 1-2-3 . . .

D oes it pay to have your soil tested? How often should soil be tested? Which lab or company is best? Are tests of all the trace elements and the CEC better? Which testing methods give the most accurate results—or the best fertilizer recommendations? And what about plant tissue analysis—does it tell anything? How can you interpret test results? What do they really mean?

For those who are not soil chemists, soil and tissue testing can be a pretty bewildering "hocus pocus" operation, so most farmers "leave it to the experts" and follow their advice religiously. But shouldn't those who make their living from the soil know more about what's going on and why, so as to make more intelligent decisions? The main thing which forms the basis for millions of dollars worth of fertilizer application is soil testing. Let's learn more about it.

We tend to get the impression, from the neat figures on the computer print-out, accurate to one or two decimal places, that soil testing and fertilizer recommendation is an exact science. After all, doesn't the soil testing lab use multi-thousand dollar lab equipment—probably a *flame spectrophotometer* (whatever that is)—plus a COMPUTER?! It's pretty impressive, isn't it?

If you request it, the soil testing lab will probably send you a detailed technical explanation of their testing methods, listing the exact procedure followed, and test reagents and instruments used. That's also impressive.

But did you know that there are different soil testing philosophies, or theories, very different methods, different ways of reporting the results, and different fertilizer recommendations that result? In the 1980s the Rodale

Research Center sent identical soil samples to 69 private and university soil testing laboratories. The results were astounding. For example, the fertilizer nitrogen recommendations ranged from zero to 230 pounds per acre, and phosphorus recommendations from zero to 150 pounds per acre! In terms of fertilizer dollars, those recommendations mean a difference of many dollars per acre. Rodale Press charged that soil testing laboratories "are largely unreliable in their recommendations," and "the unreliability of the labs is costing farmers money—conservative estimates put the figures at hundreds of millions."

Iowa State University soil scientist Alfred Blackmer has tested cornstalks from over 3,000 fields and says farmers apply too much nitrogen fertilizer, and at the wrong time (fall or early spring). Tests found that in Iowa, 75–100 lbs./acre nitrogen is the most economic rate, nearly half the usual university recommendation (*Farm Journal*, October 2004, p. 7–11).

Corn and soybean farmer and consultant Tom Novak of Sullivan, Wisconsin, grows 130 to 170 bu./acre of corn with less than 90 lbs./acre of nitrogen, but split into two applications: 8 lbs. at planting and 80 lbs. after the corn is a foot or so tall (*Farm Journal*, November 2004, p. 39).

That's pretty serious! Just by following unreliable fertilizer recommendations, farmers may be wasting millions of dollars! Let's find out what's really going on.

Objectives. Why are soil tests run, anyway? Basically, to determine the fertility of the soil, either to know how much fertilizer needs to be added or to predict the yield of a crop. Few farmers would deny that soil testing is necessary, or at least valuable.

Other methods. Actually, there are other ways of evaluating soil fertility besides chemical soil testing. For example, we can actually test-grow plants, either in a small-scale greenhouse test or in full-scale field test plots. A few farmers and many researchers do this. It takes extra time and trouble. Or we can examine growing crops for mineral deficiency symptoms, such as yellowing midvein for nitrogen deficiency and purple leaves for phosphorus deficiency. Trouble is, by the time those symptoms appear, it's about too late to do anything that year. Or, there are lab tests using certain bacteria and fungi to determine certain deficiencies in your soil. These tests aren't used much, however.

Then there are leaf and tissue tests, which determine the amounts of nutrient elements in a certain part of the plant at that time. These tests are increasing in popularity, with rapid field test kits available, besides the more detailed (and expensive) lab tests. But tissue tests, again, may be run too late to correct a deficiency that year. Also, they may not indicate a real soil deficiency. Sometimes a plant will concentrate as much phosphorus in its tissues as possible (giving a high reading on the test) just before the soil is depleted of available phosphorus *(Annual Review of Plant Physiology*, Vol. 10, 1959, p. 17). And if a plant is absorbing but not metabolizing (using) a nutrient, it will still show up in a tissue test, even though the plant has no use for it. Test interpretations seldom take nutrient balance and interactions into account.

But most farmers don't take the time or expense to have those tests run. By far the most commonly run tests are chemical analysis soil tests. Again, rapid field test kits are available, but most people send their soil samples in to a soil testing laboratory. Several tests are usually run on all samples, such as pH, nitrogen, phosphorus, potassium, calcium, magnesium, and total exchange capacity or CEC (cation exchange capacity). Other tests can be run as an option, such as percent organic matter, percent base saturation of several elements, and micronutrient (trace element) analysis.

Soil test methods. There is a multitude of different methods used by labs to test for the various elements and properties of soil. For pH, there is no problem; most labs simply mix a soil sample with a certain amount of distilled water and measure the pH with a standard pH meter. Or in an alternate method, a chemical buffer solution is slowly added to obtain a reading.

But for the other properties, exchange capacity and nutrient amounts, there are widely differing methods and philosophies governing soil testing which the farmer never hears about. The layman just assumes that because the test results from one lab show that his soil has 150 pounds of nitrogen per acre or that his soil needs 1500 pounds of lime per acre, that the tests done by another lab would turn out the same. Not so.

Most labs first dry the soil in order to standardize samples. Others use soil with its natural moisture, with the philosophy in mind that they should test under the conditions that the plants experience. Most labs then grind the soil to a uniform fineness. Some do not, realizing that plants do not have grinders. Then the soil sample is extracted; that is, it is soaked for a certain length of time in one or more extracting fluids which dissolve a certain amount of the nutrients present in the sample. Here there is wide variation. We can use water, salt solutions, alkalies, weak acids, and strong acids. Some extracting fluids work better for some nutrients or for certain types of soil. The stronger ones (or a longer extraction time) will naturally dissolve out more of the nutrients. But plant roots do not have strong acids to obtain every last microgram of a nutrient from the soil. Tests using strong extracting fluids are essentially "geology tests"—finding out the total mineral content of the soil.

After the nutrients have been extracted from the soil sample, the extracting fluid is then chemically analyzed to determine the amounts of the various nutrients it contains. This can be done by adding other chemical solutions (reagents) to it and noting certain color changes, or an expensive lab instrument can be used, either an x-ray emission spectrograph or more likely a flame spectrophotomer will be used to give "more accurate results."

The results. So, by whatever method is used, the laboratory comes up with some numerical figures of the amounts of nutrients measured. How accurate they are depends on the lab apparatus and the skill of the lab technician. But how they are reported and the units used can vary widely. Some labs may give pounds per acre (for either the upper 6 inches or 12 inches), others give parts per million, and others milligram equivalents per 100 grams of soil. And then some use elemental terms—phosphorus, potassium, nitrogen—while others use ionic or oxide forms—P_2O_5, PO_4, K_2O, NO_3, NH_4 And what exactly are they measuring? Usually it is called *available* nutrients, but what does that mean? Is it water soluble, exchangeable, extractable, or what? Available to whom—the chemist or the plant? No wonder the farmer gets confused!

Interpretation and recommendations. We have found out how much nutrients a particular sample of soil contains. So what? What does it really mean? How does it relate to growing crops and applying fertilizer? In order for soil analysis results to do the farmer any good, they must be interpreted. And here again there are many pitfalls and much variation in philosophies and practices of different labs.

The overall basic philosophy of growing crops is that the plants have certain mineral needs or requirements and that they take a certain amount of minerals out of the soil during a growing season. If the soil is deficient in one or more nutrients, they will be limiting factors in plant growth. Any available nutrients removed by plants must be replaced if they are in short supply. So far, so good. But there are additional things to consider. First, soils and plant growth needs change during the growing season. A plant's need for most nutrients is greatest during the middle and later part of its life, when most growth occurs, plus seed or fruit development. Also, nutrients can become available during the growing season and over the winter due to chemical and biological activity in the soil. Then too, the weather conditions during the year affect these variables considerably. Virtually all soil testing operates on the simplistic theory of merely replacing the nutrients removed by the crop or fertilizing to obtain a certain yield. These ideas are just not up-to-date.

Got a Hole in Your Bucket?

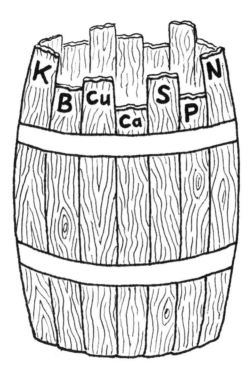

One of the basic principles of fertilizing crops is the concept of the *limiting factor,* first stated by the 19th century German chemist, Justus von Liebig. We know that plant growth and production is controlled by many factors, including light, temperature, water, carbon dioxide and the 14 nutrient elements that are usually supplied by the soil (nitrogen, phosphorus, potassium, sulfur, calcium, boron, etc.).

Each type of plant needs a certain level or amount of each factor for optimal growth and health. But in the farm environment, some factors will be in excess and others in less than ideal amounts. The factor that is at the lowest sub-optimal level is called the limiting factor, and it, more than any other factor, will limit growth. It is like trying to carry water in a bucket with a hole halfway down. Or, a better visual aid is to imagine a wooden barrel with staves of different lengths, each one representing the level of a growth factor. The barrel will only hold as much water as the level of the shortest stave, perhaps calcium.

This concept of limiting factors is another way of saying that a crop needs a certain *balance* of nutrients, more of some, less of others, and very little of others (trace elements). This is a major reason for getting a complete soil test of most of the elements needed by crops—to determine your limiting factors. Perhaps your soil is lowest in plant-available calcium, but also very low in phosphorus and boron. Those three nutrients should be the ones you need to work on the hardest. Of course you need to take into account such variables as crops expected to be grown and any special nutrient needs they have. Also, some fertilizer nutrients readily leach downward (including nitrate nitrogen, sulfur, calcium and boron) and will need to be replenished more frequently.

Adding more of the nutrients your soil already has plenty of is a waste of money and may even throw the system more out of balance and reduce crop growth. Identify and fix your limiting factors first. They are what's holding you back.

Then too, there are different goals and philosophies that the farmer can follow. Much of the time, a farmer applies fertilizer to achieve a certain yield goal, such as 150 bushel corn, or 50 bushel soybeans. Also, one can be on a maintenance fertilizer program, seeking to stay at a certain level of fertility, or one can be on a corrective or building program, seeking to improve the soil. All of these factors greatly influence the interpretation of test results and fertilizer recommendations.

But there is still one more thing—the amounts of nutrients found in a soil test must be correlated to actual crop response in the fields. In other words, test results have to be calibrated.

Calibration. This is done by conducting plant growth tests, both under greenhouse and field conditions. Different soil types must be tested, different hybrid varieties, and in different climates and latitudes. From the results, a graph can be plotted (usually of yield) going from low crop response at low rates of fertilizer to maximum response at very high rates. Figuring in fertilizer costs, a most profitable fertilizer rate is determined, which will be at a lower fertilizer rate than the maximum response, since response decreases at higher fertilizer rates. This extremely important activity takes much of the time and expense of university and industry experimental farms. It is a crucial part of the whole fertilizer-soil test procedure, because this is the major factor upon which fertilizer recommendations are made. Yet agronomists and researchers often

No Foundation

Could the foundation of the fertilizer recommendations of modern agriculture be non-existent? Yes, believe it or not. Virtually all university, extension, and fertilizer company recommendations are based on the balance-sheet theory, first set forth by the German chemist, Justus von Liebig, in 1840. His theory, arrived at after studying the growth of plants and analyzing the soil they grew in, states that plant growth and yields are in "exact proportion" to the amount of mineral nutrients they take out of the soil.

But von Liebig did not take in to account living organisms in the soil; he assumed that soils are like lifeless storage bins of water and inorganic mineral elements (*Soil*, U.S.D.A. Yearbook of Agriculture 1957, p. 3–4).

The balance-sheet theory won the day, partly due to von Liebig's prestige and argumentative skill, and it today so permeates the thinking of the "experts" that they often minimize the overwhelming importance of organic matter and living soil organisms. The experts' figures and recommendations of how much fertilizer each crop "requires" (such as 115 lbs./acre nitrogen, 40 lbs./acre phosphate, and 116 lbs./acre potash for 80 bushel corn) follow the balance-sheet theory right down the line. In other words, put in what the crop takes out.

Yet von Liebig's balance-sheet theory has definitely been proven false (*Soil*, U.S.D.A. Yearbook of Agriculture 1957, p. 7); von Liebig himself admitted it was false before his death.

How much simpler and cheaper it is to let nature's volunteer army go to work and supply crops with the kind and balance of nutrients they need?

Exchange Capacity

Plant nutrients in the soil solution (soil water with its dissolved chemicals and gases) are in ionized form, with either a positive or negative electrical charge. Clay and organic colloid particles have a negative charge on their surfaces, and hold (adsorb) positively charged ions (cations, such as NH_4^+, Mg^{++}). Some ions are held tighter than others, thus they are not as leachable by water.

In the root zone, nutrient cations can exchange places with other cations, such as hydrogen ions (these are what cause soil acidity), which come from plant roots, from the respiration of microorganisms, and from decay of organic matter. (Anion exchange can also occur, but most anions, such as NO_3^-, are directly absorbed from the soil solution.)

Cation exchange capacity (CEC), or base exchange capacity, is a measure of the cation holding ability of clay or humus. It is usually measured by soil testing laboratories. Another related measure is base saturation, which is the percent of CEC saturated by cations; that is, how much of the soil's capacity is being used. Typical CECs range from less than 5 milliequivalents/100 g for sandy soils, to 10–25 for loams to 30 or more for clays. Higher CEC soils can hold more nutrients.

While these measures have their value, they also have their limitations. For example, the soil is constantly changing, and the addition of organic matter can change the CEC. Also, in standard soil tests, not all cations are measured, so the CEC is only a part of the total picture.

admit that the correlation between field tests and soil test fertilizer recommendations is not perfect and needs much additional research.

So we see that one of the main pillars supporting the structure of soil testing and fertilizer recommendations is rather shaky. Again, the actual fertilizer/lime recommendations are based on many possible variables—field tests, soil types, soil test results, maintenance versus corrective goals, and the basic approach and philosophy of the lab's director(s) toward soil fertility and soil management.

Philosophy. This last factor is a "whole new ball game!" But it is another of the pillars holding up the structure, perhaps the very most important one. As we have hinted earlier, some people believe testing and recommendations should be based on conditions as nearly as possible approaching those found in nature, where the plants actually grow. They believe that there are natural biological systems and laws operating that will produce optimal crop growth and yields with a minimum of human interference. At the opposite extreme are people who believe that modern science and technology will provide the answers, and that growth and yields are vastly improved by the use of synthetic fertilizers, herbicides, and pesticides. Then there are others between these two extremes.

So do you see how one's overall philosophy toward soil management can dictate which and how much fertilizer to recommend? Some commonly used fertilizer materials are actually toxic to seedlings, roots, and beneficial soil organisms, not to mention herbicides and pesticides. Some destroy soil humus.

Presentation of results. The soil test results are "translated" into fertilizer and lime recommendations, generally by a mathematical calculation based on previous field test calibrations, yield goal, maintenance versus corrective goals, and overall soil management philosophy. Most labs use a computer to do the calculations and make the print-out to save time (and impress the customer?). Many labs simplify the interpretation of test results by grouping the numerical readings into several categories or levels of soil nutrients, such as high-medium-low or very high-high-medium-low-very low. Others do this by drawing a graph of nutrient levels. These practices may be helpful, but if a nutrient level is close to a cut-off line, it could be misinterpreted (for example, a level just above the "low" category).

What else? But most labs do not include in their standardized formula for calculating fertilizer recommendations such important factors as the past and future crops grown and planned for the field, and the present organic matter or humus in the soil. For example, most labs recommend adding from 1.0 to 1.25 pounds of "nitrogen" for each bushel of corn expected. They ignore the fact that if soil contains abundant humus and is healthy and alive with beneficial microorganisms, organic matter will break down during the growing season to supply over 50 pounds of nitrogen per acre. Also, if legumes have been grown on healthy soil, nitrogen-fixing root nodules will add considerable nitrogen (about 50–200 pounds per acre). Even lightning and rain add a surprising amount of nitrogen to soil, about 15 pounds per acre. Very few labs test for the two kinds of nitrogen, nitrate and ammoniacal. The growth of many crops is very different under fertilization with one or the other. And don't forget that the lab recommendations are based on standard formulas and field calibration tests. How were those tests run? What fertilizer materials, herbicides, and pesticides were used? Was the soil full of living microorganisms or sterile and dead? Was it aerobic or anaerobic? Will *your* fields behave the same way? These factors play a tremendous part in determining the yield and quality of crops.

What then? Does this mean that soil tests are totally unreliable and worthless? Absolutely not. Soil tests are a tool, an aid in (1) finding out where your soil is now, and (2) planning for the future. But they have their limitations and they do not consider all the important factors that influence crop growth and yield. They need to be used intelligently and cautiously.

How often? How often should your soil be tested? Most labs or agricultural consultants recommend testing at most once a year—or even 2, 3, 4, or 6 years. And most farmers certainly don't test their soil very often. They're just too busy trying to keep up. But think about this: soil conditions and nutrient levels change considerably during a growing season, even from week to week and day to day. Different crops and fertilizers affect soil conditions. Tillage methods and weather do too. Soil pH can change by at least two points (100 times more acidity) during a growing season. Sometimes nutritive deficiencies can be corrected during that growing season if detected soon enough.

Some farmers, especially those growing high-value crops, have their soil tested more than once a year: (1) before planting to plan fertilizer application for the year, (2) during the growing season to detect possible problems, and (3) in the fall to see what nutrients have been removed and to plan for next year. If you are trying to make big soil changes, more frequent testing the first few years may be worthwhile, then tapering off to every few years. You have to balance cost of tests *vs.* value of crops. In

general, testing is money well spent since it may save you many dollars in inputs.

Which tests? Is it necessary to have all the kinds of tests run on your soil every time, including trace elements? No, in fact most of the trace element tests are not very reliable for predicting crop response, anyway. It is most important to correct the amounts and balances of the major elements first, namely calcium, phosphorus, potassium, and nitrogen, both nitrate and ammoniacal. After these are pretty much in line, then it will be worthwhile to check trace elements, unless of course obvious trace element deficiency symptoms are visible in plants or you have a problem type of soil that is known to have trace element deficiencies.

Many soil testing labs, especially those connected to fertilizer companies and co-ops, usually do "P, K and pH" tests for phosphorus, potassium and acidity. These tell you only a little, but not nearly enough. Usually, to make basic fertilizer and lime decisions, you need those plus calcium magnesium, sulfur, organic matter and cation exchange capacity.

Sampling. Soil test results are only as good as the samples being tested. Since only a very small amount of soil from a large field is actually being analyzed, one should be very careful that truly representative samples are taken. If the soil test lab you have chosen to use has any specific instructions, be sure to follow them. In general, a composite sample is taken from a field; that is, several small samples from different parts of the field are mixed together to give an average or representative sample. In most cases this works well, but if a field has great differences in soil type, contour, or past treatment (crops, fertilizers, lime, manure), separate samples should be taken. Usually, samples should be taken from the plow layer, 6–10 inches in depth, or from the upper 6 inches if a non-plowing tillage system is used. Avoid sampling in a fertilizer band, in poorly drained areas, dead furrows, corners, or any other areas that are not representative. Stay at least 50 feet away from roads, fence rows, and barnyards.

Any tool can be used to dig down. Standard soil probes (tubes) and augers work well, but a shovel will do. Do not touch the samples with rusty iron tools or your fingers. Many (10–30) small samples from all over the field should be taken (you can use a zig-zag path as you cover the field) and mixed together to give a composite sample of 1/2 to 1 pound (1 to 2 cups). Place each composite sample in a separate paper or plastic bag with a clear label giving your name, address, and the sample designation (number or letter, such as Field A, or Field 3. Designate different samples within a field, such as A-1, A-2, A-3).

It is very important that you make a map of your farm and show where the samples were taken in each field. For consistent long-term testing, samples should be taken from the same area each year and at about the same time of year. Keep accurate records of the history of each field: crops grown, fertilizers applied (formula and amount), lime, manure, and other additions (herbicides, pesticides). Also, a record of problems and yields would be helpful.

Beware. Should you really trust your soil, crops, and precious money to soil testing-fertilizer recommendation procedures that may be unreliable, or based on questionable philosophies and field tests?

The ideal situation would be for the farmer to be knowledgeable in soil chemistry, plant physiology, fertilizers, and all the methods and procedures used by soil test labs. But such a person is a rarity. The farmer must to some extent rely on the expertise of others. However, he should certainly be careful.

Chapter 10

Help or Hindrance?
WHAT FERTILIZERS DO

T he average person—and most farmers—thinks that anything you put on your soil which is called a fertilizer must be good, must make your soil more fertile *and* must produce better crops. That couldn't be farther from the truth! As we have mentioned in previous chapters, some commonly used fertilizers and soil amendments can actually HARM the soil, can DESTROY soil structure and humus, and KILL beneficial soil organisms. You can't afford to let that happen! You need to know what you are putting on your soil and what it does.

It's elementary. In order to live and carry on photosynthesis, plants need at least 17 chemical elements. Three of them—carbon, hydrogen, and oxygen—are obtained from water (H_2O) and the carbon dioxide (CO_2) in the air. Except for the nitrogen obtained from the air by the nitrogen-fixing bacteria in legume root nodules, plants obtain most of their needed nutrients from the soil (there is some evidence that mineral nutrients can be taken from dust particles in the air; also trace amounts of ammonia and sulfur dioxide gas in the air can be used by some plants).

The three elements, carbon, hydrogen, and oxygen, partly are used as the main building blocks of most of the carbohydrate, fat, and protein molecules that make up the solid parts of cells. The basic formula for photosynthesis is a combination of carbon dioxide and water to form sugar: $CO_2 + H_2O \rightarrow C_6H_{12}O_6$. Also, the most abundant component of cells is water, from 80 to 90% in most plant cells.

The other 14 elements are needed by plants for various reasons, some as part of the structure of plants or their products, such as nitrogen in proteins and magnesium in chlorophyll, but others are essential in metabolic activities, in carrying energy, or in making enzymes function. A summary of the functions of these 14 mineral elements is found in the accompanying table (p.151).

Plants mainly use two forms of nitrogen, nitrate (NO_3^-) and ammonium (NH_4^+). These two ionic nitrogen carriers behave very differently in the soil, nitrate being easily leached into the groundwater, and ammonium being held on soil colloids and thus not easily lost. They also can have very different effects on plant growth. In many plants, fertilization by nitrate nitrogen produces rapid vegetative growth (stems and leaves); in fact, too much nitrate nitrogen can cause the plant to take up so much that it can cause nitrate poisoning in animals eating the crop. On the other hand, ammoniacal nitrogen promotes flower and seed or fruit development in many crops.

Only a relatively small amount of mineral nutrients is needed by plants. From 80–90% of the plant is water. Of the remaining 10–20% (the dry weight), carbon and oxygen make up nearly 90%. So only about 1–2% (some books say 5%) of a plant consists of mineral nutrients. But in spite of their small amount, they are absolutely necessary. If they are absent or deficient, the plant will die or be unhealthy. If a mineral deficiency is extreme, visible symptoms can be seen in the appearance of the leaves or other parts of the plant, but slight deficiencies are harder to detect. They may only be detectible as a reduced growth rate or sugar production rate, or as an amino acid imbalance, while the plant may appear to be green and healthy. At the opposite extreme, too much of some elements may be harmful or toxic. These include boron, sodium, chlorine, manganese, aluminum, iron, and copper.

Why fertilize? Since plants use mineral nutrients from the soil and are then harvested and not completely returned to the soil, and because some nutrients leach out of the soil or become chemically tied up and unavailable, the farmer must add some fertilizing materials to supplement or correct the soil's fertility and/or improve the soil's structure and pH. Other reasons for amending the soil in some cases are the correction of salinity and alkalinity (mostly in the western U.S.), helping soil to warm up faster in the spring, and overcoming waterlogging problems.

Types of fertilizer. We run into some difficulty when we try to classify fertilizers and soil amendments, because words have different meanings to different people. To a chemist, the word *organic* means any chemical substance containing carbon atoms, whether or not it is a natural or

Functions of Mineral Nutrients in Plants
(compiled from various sources)

Nutrient	Part of Enzymes	Protein Synthesis	Part of Proteins	Photosynthesis	Other Metabolic Functions	Cell Division	Part of Nucleic Acids	Part of Cell Structures	
N (nitrogen)		X	X		X		X	X chlorophyll	part of vitamins, purines, alkaloids, & metal complexes in enzyme functions
P (phosphorus)	X			X	X		X	X membranes	part of metabolic energy carriers, vit. B₁; disease resistance
K (potassium)	X	X			X				water & pH balance; phloem transport; controls leaf movement & stomata
Ca (calcium)	X	X				X		X cell wall, membranes	
Mg (magnesium)	X	X (or Co)		X				X chlorophyll, ribosomes	aids stability of nucleic acids & membranes
S (sulfur)	X	X	X		X				
Fe (iron)	X			X	X				needed for chlorophyll synthesis
Cu (copper)	X			X	X				
Mn (manganese)	X			X	X				
Zn (zinc)	X								
Co (cobalt)	X	X (or Mg)							needed for N-fixing by bacteria
B (boron)					X	X			phloem transport
Mo (molybdenum)	X				X				needed for N-fixing by bacteria
Cl (chlorine)				X	X				water balance; part of alkaloids & indoles

man-made chemical, and *inorganic* means non-carbon substances. But the original meanings of those terms were: substances produced by living organisms are organic and all other materials are inorganic. The latter meanings are useful to us here. Two other terms that can be confusing are natural (any substance found in nature, not produced by man) and synthetic (man-made, artificial). Some people often use the word chemical to refer to synthetic fertilizers, but this is not good because the term simply refers to any distinct kind of matter, whether natural or synthetic.

With regard to fertilizers and soil amendments, we may divide them into those derived from organisms, the organic materials, such as animal manures, sludge, bone meal, tankage, green manures, composts, and humic substances; and those which were not produced by organisms, the inorganic fertilizers. We can then divide the inorganics into those that are natural, such as mineral deposits (limestone, rock phosphate, potash, gypsum, greensand, granite, and fly ash); and those that are synthetic, including anhydrous ammonia, and the many "salt" fertilizers (chemically composed of a positive and a negative ion). We run into some difficulty here, because some salt fertilizers, including muriate of potash, are not man-made, but are mined from ancient salt deposits. And urea, which is found naturally in animal urine, is manufactured synthetically for fertilizer.

Organics. Still, these groupings are useful because their members share certain characteristics important to agriculture. Organic fertilizer materials contain a high proportion of complex substances (proteins, carbohydrates, fats, organic acids, vitamins, phenols, etc.), and when acted upon by living soil organisms, are slowly broken down to release both simple and complex plant nutrients; for contrary to the common belief of the "experts," plants not only absorb simple ions, but also can use more complex molecules, such as amino acids, amines, phenol, and indole. The nutrients in organic materials are very resistant to loss by leaching since they are either large molecules or are temporarily incorporated in the bodies of soil organisms. Organic substances generally provide plants with a variety of nutrients, not just one or two as in many inorganic materials. Organics will mainly be covered in Chapter 13.

Inorganics. Most of the natural inorganic materials (rock deposits) have the advantage of breaking down slowly, and they usually supply plants with a number of trace elements in addition to their main constituents. On the other hand, synthetic "salt" fertilizers are generally rapidly soluble (available to the plant) pure chemical substances (without trace elements). They quickly separate into their two ion components, some of which can be harmful to soil and soil life. Therefore, care must be taken in using the salt fertilizers to use those that cause the least harm and to use them properly, for if used in too great a concentration, salt damage can occur ("burning" of crops or seedlings), or the balance of soil nutrients can be disturbed.

Fertilizer application. There are many factors to take into consideration when applying fertilizers and soil amendments, and we cannot discuss all of them in detail here. But in general, we should consider the characteristics of the fertilizer material, the soil type and soil test results, topography of the land, climate, past fertilization and crop history, intended crop, use of herbicides and pesticides, time of application, liquid vs. dry fertilizer, cost, and whether you are on a fertility maintenance or building program. Leaving the organics for Chapter 13, we will give some general principles for most effective use of inorganic fertilizers and soil amendments. They are all based on common sense, so if you understand some of the characteristics of fertilizer materials, how they function, how the plant functions, and the principles of ecology and the soil, you can figure out for yourself the best methods. Remember that the nutrients must be made available to the plant, in the right amounts, in the right forms, and at the right time.

If you can afford a soil building program, then broadcasting the right organic and rock fertilizers will give excellent results after a year or two (they are in general slow-release fertilizers). After the first, perhaps large, application, only small amounts may be needed in later years. If your budget is tight, it would be best to give a full soil building treatment to a small acreage of a high priority crop (a high value cash crop or one that you feed to your animals), since a halfway application over a large acreage may not give visible results the first year. Then use the rest of your money on a maintenance program for your other acres. Row application, split applications, and foliar feeding can produce excellent crops a year at a time for a very low fertilizer cost per acre. Shopping around for the best buy, pooling orders with neighbors to receive bulk discounts, and doing your own blending are other ways to save money.

Crop. High yield or high value crops will give more cost-effective use of fertilizers. Some crops respond to fertilization more than others (corn responds more than soybeans). Crops you intend to feed to your livestock should receive priority over those grown to sell. Different crops have different requirements: corn needs high nitrogen, while legumes do not; alfalfa needs high calcium.

Fertilizer Fate

The average person might assume that if you put a pound of nitrogen or potassium (as part of a fertilizer material) on a field, the growing crop will take up all of that pound that year. A standard soils textbook (*The Nature and Properties of Soils,* by N.C. Brady and R.R. Weil, 1996) calls that idea a "myth." The authors say, "the reality is that nutrients added by normal application of fertilizers, whether organic or inorganic, are incorporated into the complex soil nutrients cycles, and that relatively little of the fertilizer nutrient (from 10 to 60%) actually winds up in the plant . . . during the year of fertilization." Crops use more of the added potassium, less of nitrogen and least of fertilizer phosphorus.

There are many variables that determine what happens to the fertilizer which farmers optimistically put on their fields.

1. *Type of fertilizer.* As we will see in the next chapter, there are many different chemical forms that the plant nutrient elements can take, including oxides, sulfates, nitrates, carbonates, phosphates, chelates, etc. They react differently. Some are very water-soluble, others almost insoluble. Soil pH matters a lot, as well as the presence or absence of other nutrients. Nutrients that become attached to soil particles resist leaching, although they may be so tightly attached that plants have trouble absorbing them.

2. *Soil type.* The soil's percent of clay, silt and sand, as well as exact type of clay and the pH, water content, organic matter and microorganisms, all play a role in what happens to applied fertilizers. Soils such as very sandy, acidic and drought-ridden soils are especially difficult to fertilize.

3. *Crop.* Some plant species' roots are more efficient than others at absorbing nutrients. For example, broadleafs, especially legumes, take up calcium readily, while grasses are better at absorbing potassium. Plants with small root systems respond better to added fertilizer than those with many or deep roots. Another factor is the presence of microorganisms near the root which aid in nutrient uptake, especially the fungi called mycorrhizae (see Chapter 3).

So, what happens to all the fertilizer you apply that isn't used by the crop that year? Depending on the chemical nature of the nutrient, some of it may escape into the air (nitrogen gases), leach into the groundwater, or be washed off the field in storm run-off. Those nutrients are totally lost, although they do go somewhere, probably to cause mischief there (nitrates in drinking water or phosphorus causing algae blooms in lakes, for example).

Most of the rest of unused fertilizer goes into the soil, basically into storage, but not always quickly available to future crops. Some become attached to clay particles (some ammonium nitrogen, most phosphorus, much potassium), some are held by organic matter (much nitrogen, some phosphorus, most sulfur), and some are temporarily tied up by being used in the metabolism of soil organisms. When they die, nutrients may be (or eventually will be) available to crops. See the box in the next chapter for more details on phosphorus.

Soil. Level, well drained loamy or clay soils, and soils with adequate organic matter can hold nutrients best and respond well to fertilization. Poorly drained, waterlogged, or drought-prone soils do not respond well. High sand content soils cannot hold nutrients well and require more fertilization more often. Very low (below 6.0) or high (above 7.5) pH reduces availability of several nutrients, especially phosphorus, except in very high organic matter soils, where pHs of 5.0–5.5 give best overall availability. Dry fertilizers will not be effective in dry soil. Fertilizer materials that cause a great change in pH should be avoided, or else applied long before planting or placed far from the seed.

Placement. Broadcasting can be done quickly and with low labor costs, and is good for a fertility building program, but for high-priced fertilizers, row application gives excellent results at a much lower cost. Also, row application does not nourish weeds between the rows. For row application, the fertilizer band should be placed below or to one side of the seed to prevent "burn" of seedlings. Leachable fertilizers should not be placed very deep; side dressing of nitrate nitrogen is effective.

Timing. Since seedlings do not require much nutrients, a more efficient use of fertilizers can be made by using small split applications throughout the growing season ("spoon feeding"), especially of nutrients that leach readily, such as nitrate nitrogen.

If mid-season fertilization is needed, side dressing (perhaps split into two or three small applications) and/or foliar applications can make the difference between a moderate yield and a great crop. In side dressing, a small amount of fertilizer (perhaps 100 to 200 lbs. per acre) is applied between the rows on the surface. A convenient way to do this is to mount fertilizer boxes on your cultivator for dry fertilizers. Even dry fertilizers on the surface will dissolve from rain or dew and become available to roots just under the surface. Spraying of liquid fertilizers would also be effective. If a crop is too tall to drive an ordinary applicator through, a fogger, a "highboy" or an airplane with foliar application can be used, or fertilizer can be applied in irrigation water. Fall application of fertilizers that break down slowly (rock fertilizers, organic matter) is best for the next year's crop.

Salts. Synthetic fertilizers with a low "salt index" are preferred, in order to prevent salt damage to seedlings, roots, and soil life. The salt index is a measure of the fertilizer's ability to draw water out of living cells by osmosis. The salt index of some common fertilizers are (from *Farm Chemicals Handbook*, 1981, p. B74):

Material	Salt Index
anhydrous ammonia	47.1
urea	75.4
ammonium nitrate	104.7
ammonium sulfate	69.0
sodium nitrate	100.0
potassium nitrate	73.6
calcium nitrate	52.5
mono-ammonium phosphate	29.9
diammonium phosphate	34.2
superphosphate	7.8
triple superphosphate	10.1
potassium chloride	116.3
potassium sulfate	46.1
sulfate of potash-magnesia	43.2
calcitic limestone	4.7
dolomitic limestone	0.8
calcium sulfate	8.1

Note the much higher salt index of potassium chloride (muriate of potash), compared to potassium sulfate. This, besides the chlorine content of muriate of potash, is another good reason to use potassium sulfate (0-0-50) instead of muriate of potash (0-0-60).

Nutrient balance. Crops grow better and produce higher quality food if the soil nutrients are in a relative balance. The exact balance will differ slightly with different crops, but for most crops, the proportion of calcium should be higher than potassium. This will produce what is sometimes called proteinaceous crops, whereas too much potassium produces "carbonaceous" crops, high in carbohydrates and bulk (cellulose), but not nutritionally complete. Also, if phosphorus is low relative to calcium, animals or humans who eat the plants may suffer reproductive problems (from F.A. Gilbert, *Mineral Nutrition of Plants and Animals*, 1948). If there is too much magnesium relative to calcium, unhealthy plant cell walls may be formed. There is a host of other nutrient interactions that have been discovered; see the accompanying table (p. 158).

Nutrient form. There are two main types of nitrogen used by plants, nitrate and ammonium. In some cases, nitrate nitrogen tends to produce lush vegetative growth, while ammonium nitrogen produces flowers and seed if the plant is mature enough. So, if you are growing alfalfa for hay, use a nitrogen source that contains nitrate, but if you want grain, or alfalfa for seed, use an ammonium-containing source.

Potassium. Most agronomists or fertilizer experts tend to recommend too much potassium in relation to the other major nutrients. Some potassium is necessary, but if phosphorus and calcium levels are too low compared to potassium, low quality, fibrous crops result, lacking high quality proteins and other nutrients (crops may grow well and look good, but quality is still poor). The same thing is true of magnesium. Some is needed, but not too much in relation to calcium. Calcium and phosphorus "build quality" into a crop. Unfortunately the "sufficient" levels of them on soil tests are usually too low.

Trace elements. Some people worry about trace element deficiencies, and commercial fertilizers are available with trace element "packages." Trace element availability is affected by soil type, pH, and the balance of other nutrients, especially the major nutrients (see the nutrient interaction table). Except in cases where a severe trace element deficiency is obvious, it is best to first get the major nutrients in balance. Usually that will solve the trace element problem; if not, then it can be attacked. The trace element packages offered by fertilizer companies are just a shotgun approach and mostly a waste of money.

Nutrient Interactions

If the element (or soil condition) at the left is too high (H) or too low (L), a deficiency of the element at the top can result. (Information compiled from many books and journals.)

	N	P	K	Ca	Mg	S	Fe	Al	Cu	B	Zn	Mo	Mn	Na	Cl	Co
N (nitrogen)																
P (phosphorus)	L		H				H		very H	H	H	H				
K (potassium)		H		H	H					H	L					
Ca (calcium)		H	H		H		H			H	very H					
Mg (magnesium)		H	H	H						H						
S (sulfur)	L															
Fe (iron)		H + Al low pH	H + Al, Mn + low pH	H + Al, Mn + low pH					H + Mn				L			
Al (aluminum)		H + Fe low pH	H + Fe, Mn + low pH	H + Fe, Mn + low pH					H							
Cu (copper)							H					H				
B (boron)			H	L	H											
Zn (zinc)																
Mo (molybdenum)	L						H		H							
Mn (manganese)	H		H + Fe, Al + low pH	H + Fe, Al + low pH			H + low pH		H + Fe							
Na (sodium)		H	H	H												
Cl (chlorine)		H	H													
Co (cobalt)																
Ni (nickel)							H or L + Mn									H
pH	L + H Fe, Al	L + Al Fe, Mn		L + Al Fe, Mn	L + Al Fe, Mn		H or L + Mn		H	H or L	H	L	H + org. m.			
organic matter									H	L	H	L	H + H pH			

Foliar Feeding

Plants are able to take in many nutrients through their leaves. Spraying materials on the above-ground part of the crop is often called foliar feeding. It is not practical to supply a plant's total needs of the major elements in this way, but foliar feeding is an excellent way to supply secondary or trace elements, to "energize" a crop whose growth is lagging, and to "switch" a crop from vegetative growth to flower and seed production with ammoniacal nitrogen. And it can be done for only pennies per acre of materials in some cases.

Of course you need special spraying equipment (or to hire custom or aerial spraying). The finer the spray the more effective it is in being absorbed by the plants and the cheaper the cost of materials (less is needed). An ordinary field sprayer does little good. Sprayers that produce a mist or fog are better, and one that homogenizes is best. A highboy tractor allows tall crops such as corn to be sprayed late in the growing season (foliar spraying is often required every 10 days or two weeks).

If foliar feeding is done to supply deficient trace elements, results may or may not be satisfactory. Generally, unless the major and secondary elements are adequate and balanced in the soil, trace element application may do little good. Trace elements come in various chemical forms, both inorganic and organic. Some are better than others, and some are very expensive. However, only very small amounts of trace elements are needed by plants. You should never add trace elements just because you think they *might* be deficient, since too much of some trace elements can be toxic. A soil test should be done and only the recommended amount should be applied. "Shotgun" mixtures of trace elements can be harmful if one or more elements are not deficient. Often trace element deficiencies will magically disappear after you build up soil humus and organisms, since these supply trace elements and break them out of soil minerals.

You can buy pre-mixed foliar sprays, but it is cheaper to mix your own from scratch. The mixing and use of foliar sprays can get very complicated, with special mixes for each crop and situation. We cannot go into that much detail here, but there are some simple but effective foliar sprays that give good results with most crops. When mixing and using foliar sprays, use only clean equipment that has never been used for herbicides and insecticides, and use chlorine-free ingredients and water (or at least no more than 1% chlorine). Soft or deionized water is best; minerals in hard water may interfere with spray ingredients.

A good beginning spray is liquid fish emulsion, diluted at 2 gallons per 100 gallons water. The fish emulsion may have lumps that would clog a sprayer, so add one gallon of fish to 10 gallons of water, mix and strain through a fine screen or cloth strainer, and then finish diluting. Spray this at a rate of 2 gallons per acre.

Another excellent formula is about 3 parts liquid fish to 1 part liquefied seaweed (about 6 quarts fish and 2 quarts seaweed per 100 gallons water). Both of these products from the sea are full of trace elements, plus the seaweed has plant hormones and growth stimulants. Fish also contains about 10% nitrogen. Be certain that these spray mixtures are not alkaline, but between pH 5.0–6.5 (inexpensive pH testing paper can sometimes be purchased from a drugstore). First adding 1 pint to 2 quarts of liquid phosphoric acid to the water will neutralize alkaline ingredients. Spray at the rate of 1 quart per acre of fish-seaweed mixture.

A more complex formula for seed crops in the middle of their growing season is: in 100 gallons water, mix in this order:

1. One to four pints liquid phosphoric acid (40–45%, available from feed stores), depending on level of phosphorus in soil.

2. One gallon household ammonia (substitute 3–4 lbs. calcium nitrate if the crop is still young).

3. If soil is low in potassium, add a potassium source, either 1/4 to 3 lbs. of potassium sulfate or 1/2 to 4 lbs. of potassium hydroxide (this is a type of lye; dissolve first in a little water).

4. Five lbs. of colloidal (soft rock) phosphate. Stir into a little water, let settle and use the milky water.

5. Any trace elements that are seriously deficient. Up to 5 lbs. of filtered poultry manure is a good source of trace elements, especially boron. Manganese is a trace element that promotes good seed or fruit development; use if needed (too much is toxic).

Most crops can use some iron; too much is not harmful. Leave some scrap iron in a bucket of water and use the rusty water as an iron source.

6. Two gallons of liquid fish emulsion or the fish-seaweed combination covered above is an excellent option to supply trace elements and growth hormones.

7. One quart emulsified oil (crop oil, dormant oil, Stoddard solvent); use no more often than once a month.

Spray the above mixture at a rate of 7 gallons per acre, or enough to wet plants. For non-seed crops (alfalfa, hay, grasses), use 3–4 lbs. of calcium nitrate in place of the ammonia and omit the potassium source. Do not use the trace element manganese. Alfalfa requires high calcium, which is best supplied by liming the soil; adding 3–5 lbs. of finely ground calcium carbonate (high calcium lime) to the spray at step 4 (above) will supply some. Do not add fish emulsion within two weeks before harvest of hay or forages, because cattle do not like the taste.

For root crops and potatoes, increase the amount of phosphoric acid and potassium source. Substitute potassium nitrate for ammonia (from 10 lbs. per 100 gallons water for carrots, to 20 lbs. for beets, up to 50 lbs. for potatoes).

Up to 5 gallons of molasses per 100 gallons water is an excellent addition for sprays (add at step 4, above). It supplies iron, stimulates soil life, and increases palatability and nutrient content of hay crops, especially if sprayed just before harvest.

The figures used above apply to the use of a homogenizing sprayer. For ordinary mist sprayers, the amount sprayed per acre may have to be doubled or more. For crops with large leaf area (soybeans, vegetables) or greater growth, amounts also will have to be increased.

The best times to spray are in foggy weather or in the early morning (4–6 a.m.), since this is when the plants take in materials best. The evening is another good time to spray, after 7 p.m.

A good way to test whether the spray mixture is going to help your crop is to spray several plants with a small hand sprayer. Wait a half hour and test the sugar content of sprayed plants with a refractometer (see Chapter 18), compared to unsprayed plants. If the sugar content increased, the mixture is good.

A spray to "switch" a seed crop from vegetative growth to flowering should contain ammonia or ammonium nitrogen rather than nitrate nitrogen. For small acreage or gardens, a spray of diluted vinegar (2 cups per gallon of water, spray 1 gallon per 100 square feet) can be used about halfway through the growing season. For larger acreage, a light side dressing (100–150 lbs per acre) of superphosphate (0-20-0) will do the trick. If these aids are used for "switching," be sure the soil has enough fertility to fill out the seed or fruit. Additional side dressing may be necessary. The same thing applies to other growth stimulants, such as seaweed, hormone, and enzyme sprays. It does little good to stimulate growth if the soil cannot supply the needed nutrients.

Dangers. To avoid or reduce stress on plants and harm to soil and soil organisms, the following materials should be avoided:

1. Ammonia or materials that release large amounts of ammonia (urea and diammonium phosphate). Large amounts of ammonia can kill roots and soil organisms, destroy humus by making it soluble, and thereby cause soil hardening. You may get by if you apply small rates.

2. Chlorine-containing fertilizers (mainly muriate of potash, also calcium chloride and ammonium chloride). Although plants need small amounts of chlorine and most can tolerate a fairly large amount (up to 100–140 parts per million), and although chlorides can leach from the soil, amounts of chloride can temporarily be found in the soil that can not only injure plants, but can kill the beneficial soil organisms. Even if crops can tolerate chloride, their quality can still be lowered. High chloride is known to reduce yield and growth of corn. Potato and tobacco farmers usually avoid chlorine-bearing fertilizers because these crops are highly sensitive to chloride. Corn, soybeans, sweet potato, cucumber, strawberry, pepper, blueberry and citrus are moderately sensitive to chloride at over 100 parts per million.

3. Dolomitic limestone or other magnesium-containing fertilizers (such as sulfate of potash-magnesia or magnesium sulfate [Epsom salts]) should not be used on soils that are not deficient in magnesium. The magnesium

may cause nitrogen to be lost into the air, tends to form a crust on the soil, and causes low quality crops if the calcium level is low.

4. High concentrations of any salt fertilizer, especially those with a higher salt index, plus large applications of animal manures or whey, can cause salt damage to roots and kill soil life.

 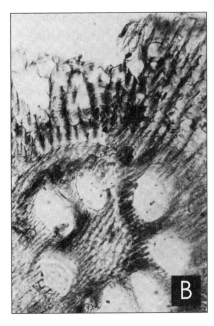

A. Corn roots grow around the zone where salt fertilizer was placed in the soil.
B. Corn root tissues damaged by salt fertilizer (cross section magnified 100 times under the microscope).

Chapter 11

Fertilizers —

WHAT'S IN THE BAG?

U nfortunately, fertilizers are still licensed, promoted, and sold under the outmoded "magical" N-P-K concept, but anyway, when you buy a bag of, say 9-23-30, DO YOU KNOW WHAT'S IN THE BAG?

Chances are, you don't. What, chemically speaking, makes up the 9 pounds of "N"? What accounts for the 23 pounds of "P"? And what about the 30 pounds of "K"?

Did you know there are hundreds of different ways to mix a 9-23-30 fertilizer? That there are dozens of *sources* of nitrogen, phosphorus, and potassium? That some types of fertilizers and soil conditioners can actually *harm* your soil and crops or cause crops to be unhealthy? That two different sources of nitrogen can behave very differently in the soil, having opposite effects on pH and on plant growth, one causing vegetative growth and the other promoting seed development? In this chapter, we want to examine some of the commonly used sources of plant nutrients and see how fertilizer formulas are figured. *You need to know what's in the bag!*

Labels. Federal and state fertilizer laws require that fertilizers be labeled to show the guaranteed nutrient content of at least the "big three"—N, P, and K—and sometimes secondary and trace elements. So your first source of information should be the label on the bag or tank.

The percent of plant nutrients (or pounds per hundred) listed on the label is called the *grade* or *analysis*. At least the three elements, nitrogen, phosphorus, and potassium, are often referred to in that order as a ratio, such as 10-5-10 or 2:1:2, although not all three may be present, as in 0-20-

0. Any additional elements can be added, but must be identified, such as 10-5-10-2S, which contains 2% sulfur.

The analysis can be given either for the P and K oxides (called phosphoric acid or P_2O_5 and potash or K_2O) or for the elemental forms, phosphorus (P) and potassium (K). Most labels will have both, but the tendency is to label on the elemental basis. One can easily be converted to the other: % or pounds of P_2O_5 multiplied by 0.44 gives P (in reverse, % or pounds of P multiplied by 2.29 gives P_2O_5), and % or pounds of K_2O multiplied by 0.83 gives K (while % or pounds of K multiplied by 1.2 gives K_2O).

In the bag? However, what you are getting in the bag or tank is NOT pure elemental nitrogen, phosphorus, or potassium, and not even P_2O_5 or K_2O. If you look carefully, the label may read P_2O_5 *equivalent* or K_2O *equivalent*. The material you buy does contain the stated percentage of nitrogen, phosphorus, and potassium, but those chemical elements are in *combined form* as chemical compounds, either as inorganic or organic compounds. They are never in the oxide forms of P_2O_5 or K_2O—those are just standards for comparison and labeling purposes. And as we said earlier, there are dozens of different commonly used sources of plant nutrients, each with its own properties and effects on soil and crops.

Physical form. Also, fertilizers can be prepared in several physical forms, which have their own specialized equipment for storage and application. These are (1) *gas*, mainly anhydrous ammonia, a source of nitrogen; (2) *dry solids* (either powders, crystals, or pellets), which can contain one or more nutrients; (3) *liquids*, one or more nutrients in water solution; and (4) *slurries* and *suspensions*, combinations of dry and liquid components. In general, there is no significant difference in the effectiveness of one form of fertilizer over another. The big differences are in the chemical *source*. So let's look at common sources of N, P, and K . . . as a "shopping list" if you will, of possible nutrients to choose from.

Nitrogen. The air around us is mostly elemental nitrogen—78%, but in this form, plants cannot use it. Plants mainly use nitrogen in the nitrate (NO_3^-) and ammonium (NH_4^+) forms. Soil organisms convert ammonia (NH_3) and ammonium forms to the nitrate form, which readily leaches into the groundwater (the conversion processes are called nitrification). Waterlogged soil can also result in nitrogen loss through a more or less opposite series of microbial changes called denitrification. Nitrogen from the air can be trapped, or fixed, by nitrogen-fixing bacteria which can either live independently in the soil or in the root nodules of legumes. That is, provided the soil environment is favorable for beneficial soil organisms, and not sterilized by toxic substances. Additional nitrogen is supplied by lightning and rain (but rain also causes leaching of nitrates).

The Nitrogen Cycle

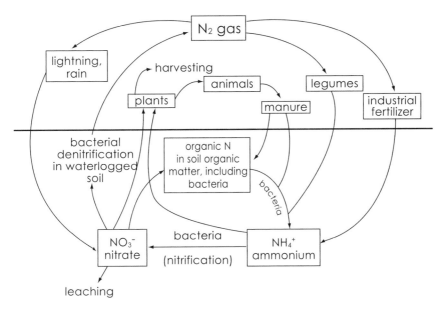

The greatest need of nitrogen in typical plants is during their time of maximum vegetative growth. Since nitrogen can be so easily lost from the soil, an effective fertilization strategy is to provide a nitrogen source or sources that are non-leachable and available over long periods of time, or to use split applications during the growing season.

A look at the accompanying table will reveal great diversity in nitrogen sources, each with its advantages and disadvantages. There is no one "best" nitrogen source. It all depends on what you need and what your local conditions are—your soil type, soil test results, crops previously grown and planned, weather, tillage methods, available machinery, and budget. Only one of the sources cannot be recommended: anhydrous ammonia. Its disadvantages of killing soil life and hardening soil by destroying organic matter far outweigh its advantages. In fact, the use of soil injection of anhydrous ammonia was perfected during World War II, when it was used to make soil hard to build landing strips for bombers (*Acres U.S.A.*, May 1981, p. 5).

Principal Nitrogen Sources
(compiled from several sources)

Material	Chemical Formula	% Nitrogen	Form	Acidity (−) or Alkalinity (+)*	Advantages, Disadvantages
ammonia	NH_3	82 (gas)	gas (anhydrous ammonia), liquid (aqua ammonia)	−148 (gas) −36 (liquid)	High N analysis; non-leachable after changes to NH_4^+ on soil colloids. Kills soil organisms; hardens soil; can be lost to air; at first causes alkalinity, then acidity.
		20-24 (liquid)			
urea	$CO(NH_2)_2$	42-46 (dry) 20 (liquid)	dry, liquid	−71	High N analysis; non-leachable after changes to NH_4^+ on soil colloids. Can change to ammonia and escape into air if applied on surface.
ammonium nitrate	NH_4NO_3	33-34	dry	−62	High N analysis; supplies two types of N. Nitrate is leachable.
ammonium sulfate	$(NH_4)_2SO_4$	20.5-21	dry	−110	Non-leachable N; also supplies sulfur. Low N analysis; medium price; causes acidity.
sodium nitrate (nitrate of soda)	$NaNO_3$	16	dry	+29	Immediately available N. Low N analysis; leachable N.
potassium nitrate	KNO_3	13.5-13.8	dry, liquid	−23	Immediately available N; also supplies potassium. Low N analysis; leachable N.
calcium nitrate (nitrate of lime)	$Ca(NO_3)_2$	15.5-17	dry, liquid	+20	Immediately available N; also supplies calcium. High price; low N analysis; leachable N.

Principal Nitrogen Sources (continued)

Material	Chemical Formula	% Nitrogen	Form	Acidity (−) or Alkalinity (+)*	Advantages, Disadvantages
mono-ammonium phosphate (MAP)	$NH_4H_2PO_4$	11-13	dry	−58	Non-leachable N; also supplies available phosphorus. Low N analysis.
diammonium phosphate (DAP)	$(NH_4)_2HPO_4$	18-21	dry	−70	Non-leachable N; also supplies available phosphorus. Can injure seedlings (high ammonia).
ammonium thiosulfate	$(NH_4)_2S_2O_3$	12-18.9	liquid	neutral	Non-leachable N; also supplies sulfur. Can injure plants if used wrongly.
animal manures	various organic compounds	0.6-2.0	dry, liquid	slightly alkaline to slightly acidic	Non-leachable N, available over long period; improves soil structure; also may supply many other elements. Low N analysis (but N content can grow with biological activity); large amounts can compete with crops, or (animal manures) can cause salt damage to roots.
green manures	various organic compounds	0.7-7	plowed under	neutral	
sewage sludge	various organic compounds	2-6.5	dry, liquid	slightly acidic	
tankage	various organic compounds	3-10	dry	slightly alkaline to slightly acidic	
bloodmeal	various organic compounds	13-15	dry	slightly acidic	

* Acidity = pounds of $CaCO_3$ needed to neutralize acid per 100 pounds of material. Alkalinity = equivalent needed to neutralize alkalinity per 100 pounds of material.

Ammonia and Soil Don't Mix

Quotes from an article by M.M. Mortland entitled "Reactions of Ammonia in Soils" in *Advances in Agronomy*, Volume X, 1958:
"The immediate effect of an injection of anhydrous ammonia or application of an aqueous solution of ammonia, is to create a very alkaline reaction in the vicinity of the application."
He then gives several examples:

original pH	pH after ammonia
6.0	8.1
5.0	6.2, 7.1
4.9	6.5, 8.0
7.3	8.8, 9.5
8.1	10.1, 10.6

"Ammonium hydroxide has a strong solvent action, as well as a hydrolytic action on organic matter. Thus applications of anhydrous ammonia or aqueous solutions of ammonia which result in localized regions of high concentration in the soil might be expected to result in solubilization and hydrolysis of certain fractions of the soil organic matter.

"Some published information indicates that ammonia may have an effect on soil structure. Smith (1954) has suggested that anhydrous ammonia is a "soil conditioner" in its effects on the physical properties of soil. His contention is that organic compounds are dissolved in aqueous solutions of ammonia, then coat the soil aggregates which become water-stable upon drying. Other results indicate the formation of a poorer soil structure. That NH_4OH has dispersive effects on colloids under certain conditions is well known. . . . It tends to coagulate soils having neutral or alkaline reaction, whereas it tends to disperse or deflocculate acid soils. . . . Such changes in properties of constituent minerals alter the physical properties of the soil.

"Ammonia, however, may be toxic to living organisms. It may exert a toxic or even lethal effect on the cells since at high concentration ammonia is a cell poison. In the application of anhydrous ammonia to soils, the tops of plants are affected wherever the free gas contacts them. Some effect on the roots when crops are side-dressed may also be expected. The extent of injury may be quite variable.

"Since ammonia is toxic to cells at high concentrations, it might be expected to affect the soil microbial population. . . . Eno and Blue (1954) found that anhydrous ammonia applications to Aredondo loamy fine sand and Lakeland fine sand decreased the number of soil fungi. Both bacteria and fungi were decreased the first day. By the tenth day fungi were still reduced in number while bacteria had increased and were 6 to 25 times as numerous as in the check."

Applying anhydrous ammonia.

Phosphorus. Soils have a super-abundance of phosphorus, 1000–2000 pounds per acre. The only trouble is, most of it is unavailable to the plant. From 40–75% is chemically "locked up" in the form of compounds with calcium, iron, aluminum, manganese and other elements. These inorganic forms of phosphorus are for all practical purposes unavailable to the plant on a short-term basis because the phosphorus is so tightly bound. High or low pH greatly reduces the availability also.

From 25–60% of the soil's phosphorus is in the form of organic compounds and part of the cells of soil organisms. By the action of soil organisms on organic and inorganic forms of phosphorus, most of the plant's supply of phosphorus is slowly made available.

Which would you buy, a bag of 0-20-0 or a bag of 0-2-0? The fertilizer label proudly tells you how much phosphorus is "in the bag" in a *water soluble form,* which is the form plants can use. But what the label doesn't tell you is that within a matter of *a few hours,* nearly all (80–90%) of the water soluble phosphorus has been chemically converted to water insoluble forms, and is thus not immediately available to your plants. You have to read technical fertilizer books to find that out. Yet most of the phosphorus sources used in commercial fertilizers have started out as insoluble rock phosphate and have been chemically treated to produce a large percent of soluble phosphate to make the label analysis look good. All of which expense (out of your pocket) is undone in a few hours!

A plant's need for phosphorus is fairly constant throughout its growth period. Needs are as high during young growth as later, but the root system may not be developed enough to supply the need. Plenty of nitrogen aids in phosphorus uptake. And don't forget about those beneficial soil fungi, the mycorrhizae, which live in and on plant roots and help in phosphorus absorption (see Chapter 3).

The accompanying table of phosphorus sources (p. 175) shows a variety of materials, with varying properties. Again, the extra expense to gain increased phosphorus solubility in the first six materials is largely wasted.

Potassium. All soils except sands have enormous amounts of potassium, but like phosphorus, most (98–99%) is unavailable, being "locked up" in soil minerals. What potassium that is available is held on clay or humus colloids, which is why sandy soils cannot hold much and must have an adequate amount added each year. Except for sandy soils, the mineral reserve should supply most potassium needs. If fertilizer potassium is added, any excess not held by soil colloids can leach away.

Phosphorus

Although plants and animals need many different mineral elements, some are needed in greater quantities than others, and phosphorus is one of those major elements, one that is often seriously deficient—in forms plants use—in today's soils. Phosphorus is essential for all life, both plant and animal. It is part of the molecules that make up genes (DNA) and those that carry genetic information from the genes to the "protein factories" of the cell (RNA). Also, phosphorus is a necessary component of several kinds of molecules that carry and temporarily store cellular energy (high energy phosphates), both during photosynthesis, when the sun's energy is trapped, and later during respiration, when food energy is released.

When a plant is deficient in phosphorus, it will continue to carry on photosynthesis for a time, but its energy production and food transfer activities will "shut down." Therefore, sugars from photosynthesis will accumulate, but will not be changed into proteins and fats. Sometimes a purple pigment (anthocyanin) will accumulate in the leaves. The reserve supply of phosphate ions in the cells will quickly be used up to form the organic phosphate carrier and genetic molecules, and once that happens, phosphorus is no longer available to new growing cells where it is vitally needed.

There are several factors that influence soil phosphorus availability to plants. Since most soils have plenty of phosphorus in chemically tied-up forms, we should pay special attention to factors that can overcome this problem. In soils of low organic matter content, lower pH (below 6.5) causes much phosphorus to become tied up in iron and aluminum compounds. However, high organic matter counteracts this and gives good P availability down to pH 5 or 6. Several studies have proved that organic matter and compost increase the availability of not only the phosphate already in the soil, but also of added rock phosphate fertilizers. Organic acids produced by microorganisms are at least partly responsible. In one experiment, application of 80 pounds per acre of soluble phosphate alone tripled the uptake of phosphorus by crops, but mixing the same amount of phosphate with organic matter (spent brewer's hops) increased uptake five times. The role of root mycorrhizae in increasing phosphorus absorption has already been covered in Chapter 3.

Optimum soil moisture, temperature, and aeration are very important for phosphorus availability because they increase microbial activity in the soil, plus roots need optimum conditions for effective nutrient uptake. In another experiment, an increase in soil temperature from 41° to 80°F increased the P uptake by corn almost four times and the yield (dry matter) over 5 times.

Besides promoting general growth and yield, phosphorus also directly or indirectly increases crop quality, drought resistance, disease resistance, nitrogen fixation, and decreases maturity time, especially in cool temperatures.

A plant's need for potassium starts out small and grows continually. In corn, about one-fourth of the potassium goes to the grain and three-fourths are in the rest of the plant.

Only several fertilizer materials are commonly used (see accompanying table, p. 177), and one of those, potassium chloride (muriate of potash, 0-0-60 or 0-0-62), which is used far more than any other, cannot be recommended because it contains 47.6% chlorine. Some crops are sensitive to chloride, especially potatoes and tobacco—also corn, soybeans, sweet potatoes, strawberries, cucumbers, peppers, citrus, and blueberries—in amounts over 100 parts per million of chloride. Muriate of potash is a high salt fertilizer (see Chapter 10). High salt concentrations injure or kill roots and interfere with the normal life activities and reproduction of beneficial soil microorganisms. Some agronomists, extension agents, and fertilizer dealers recommend levels of muriate of potash for alfalfa of 200 to 800 pounds per acre. Assuming that the plow layer of soil weighs about 2 million pounds, that's about 50–200 parts per million chloride! Is it any wonder that many alfalfa roots have no nitrogen-fixing bacterial nodules? The "experts" and the textbooks they write say that chloride forms salts with calcium, magnesium, and sodium, and quickly leaches out of the soil. One study showed that it takes two years for the chloride to leach below two feet (*Better Crops with Plant Food*, Vol. 66, Summer 1982, p. 26–28). Is

Principal Phosphorus Sources (compiled from several sources)

Material	Chemical Formula	% Phosphorus	% P_2O_5	Form	Acidity (−) or Alkalinity (+)*	Advantages, Disadvantages
superphosphate (monocalcium phosphate)	$Ca(H_2PO_4)_2$ + $CaSO_4$	7-9.6	16-22	dry	neutral	Also supplies calcium & sulfur (also contains calcium sulfate). High Price.
triple superphosphate (monocalcium phosphate)	$Ca(H_2PO_4)_2$	19.6-20.5	45-47	dry	neutral	Also supplies calcium. High Price.
mono-ammonium phosphate (MAP)	$NH_4H_2PO_4$	8.7-21	20-48	dry	-58	Also supplies non-leachable nitrogen.
diammonium phosphate (DAP)	$(NH_4)_2HPO_4$	20-23.1	46-53	dry	-70	Also supplies non-leachable nitrogen. Can injure seedlings (high ammonia).
ammonium polyphosphate	$NH_4H_2PO_4$ + $(NH_4)HP_2O_7$	15-26.6	34-61	liquid, suspension	slightly acidic	High P analysis; makes some trace elements more available (Fe, Zn, Mn); also supplies nonleachable nitrogen. More slowly available.
phosphoric acid	H_3PO_4	23-24	52-54.5	liquid	-110	High P analysis. High Price.
hard rock phosphate (calcium fluorapatite & hydroxyapatite)	$3\ Ca_3(PO)_2 \cdot CaF_2$ + $3\ Ca_3(PO_4)_2 \cdot Ca(OH)_2$	11.5-17.5	26-40	dry	slightly alkaline	Also supplies trace elements (Fe, Mg, B) and calcium. Low availability (but increases with biological activity).
soft rock phosphate (colloidal phosphate, rotten phosphate)	same as above + clay	8-11	18-25	dry	slightly alkaline	Higher availability than hard rock phosphate; also supplies trace elements (Fe, Mg, B) and calcium.
basic slag	contains $CaSiO_3$ & P_2O_5	3.5-8	8-18	dry	alkaline	Also supplies trace elements (Fe, Mg, B, Mo, Cu, Zn, Mn) and calcium.
bone meal	various organic compounds	9.7-13	22-30	dry	slightly alkaline	Also supplies nitrogen.

* Acidity = pounds of $CaCO_3$ needed to neutralize acid per 100 pounds of material. Alkalinity = equivalent needed to neutralize alkalinity per 100 pounds of material.

this "quickly"? According to the book, *Fertilizer Technology & Use*, 2nd ed., 1971, p. 331, the use of muriate of potash in the row along with the seed may cause salt injury to the seedling; on clay soil about 18 pounds per acre may be injurious, but the danger level is lower on sandy soils.

But what's in the bag? Now that we have looked at the most common sources of major nutrients, let's see how a blended fertilizer is put together and how those "magic numbers" on the label are determined.

Again, what are you actually getting in the bag or tank? Some people may think they are buying pure plant nutrients, N-P-K (nitrogen, phosphorus, potassium), but that is not the case. You are not even buying P_2O_5 or K_2O, for the label should state that there is a guaranteed amount of P_2O_5 equivalent and K_2O equivalent. You are buying some N, P, and K, but they are chemically combined with other elements. The tables of nutrient sources found earlier in this chapter give the chemical formulas of the non-organic materials. These formulas show what other elements are combined with the nitrogen, phosphorus, and potassium (refer to Appendix A for the chemical abbreviations used). The trend today is away from the confusing use of P_2O_5 and K_2O, toward giving the nutrient content in terms of the elements, N, P, and K. (other elements can also be given after those three, as long as they are identified).

Examples. Let's start with a simple case, a single fertilizer material. One example is potassium chloride (muriate of potash), which is labeled 0-0-60 or 0-0-62 using the K_2O basis. Its chemical formula is KCl, which means that it contains not only potassium, but also chlorine. Its percent of K_2O is 60–62 (49.8–51.55 elemental K) and it is labeled 0-0-60 or 0-0-62 since the chlorine is not normally labeled, although it can have detrimental biological effects in the soil.

Another example would be diammonium phosphate, which is $(NH_4)_2HPO_4$. It contains two major plant nutrients, nitrogen (18%) and phosphorus (20%). Thus its label would be 18-20-0, or if converted to P_2O_5 equivalent, 18-46-0.

Now let's look at a very commonly sold *blended* or mixed fertilizer, 9-23-30. Actually, the formulation can be mixed in many ways by using various fertilizer materials, but is usually made by mixing equal amounts of diammonium phosphate (18-46-0) and muriate of potash (0-0-60). Since these two materials are blended half-and-half, there would be half the numerical amounts per 100 pounds, so:

Principal Potassium Sources
(compiled from several sources)

Material	Chemical Formula	% Potassium	% K₂O	Form	Acidity (–) or Alkalinity (+)*	Advantages, Disadvantages
potassium chloride (muriate of potash, kalium potash)	KCl	49.8–51.5	60–62	dry, suspension	neutral	Low price. Chlorine kills soil organisms, can harm plants (contains 47.6% chlorine).
potassium sulfate (sulfate of potash)	K₂SO₄	39.8–44	48–53	dry	neutral	Nearly chlorine-free (2%); also supplies sulfur.
potassium nitrate (nitrate of potash)	KNO₃	37–38.7	44–46.6	dry, liquid	–23	Nearly chlorine-free (1%); also supplies nitrogen. High price.
sulfate of potash-magnesia (potassium magnesium sulfate, sul-po-mag)	K₂SO₄•MgSO₄ + MgCl₂ + H₂O	18.3–19.2	22–23	dry	neutral	Nearly chlorine-free (1%); also supplies magnesium & sulfur.
granite dust; greensand, glauconite	variable: potassium aluminum and iron silicates	3–5 (granite) 6–7 (greensand)	3.6–6 7.2–8.4	dry	slightly alkaline	Potassium released slowly; also supplies trace elements (Fe, Mg).
animal manures	various organic compounds	0.3–0.8	0.36–1.0	dry, liquid	slightly alkaline to slightly acidic	Potassium released slowly; also supplies non-leachable nitrogen. Large amounts can compete with crops or cause salt damage to roots.
sawdust	various organic compounds	6–7	7.2–8.4	dry	slightly acidic	Potassium released slowly. Large amounts can compete with crops.
wood ashes	various organic compounds	7–18	8.4–21.6	dry	alkaline	Also supplies trace elements. Can injure seedlings.

* Acidity = pounds of CaCO₃ needed to neutralize acid per 100 pounds of material. Alkalinity = equivalent needed to neutralize alkalinity per 100 pounds of material.

50 pounds diammonium phosphate =	9-23-0
50 pounds muriate of potash =	0-0-30
total	9-23-30

Next, a more complicated one. Here are two different commercially used formulas for a 6-24-24 fertilizer:

10.0 pounds 4-44-0	10.0 pounds 4-44-0
31.25 pounds . . . 18-46-0	21.25 pounds 18-46-0
11.5 pounds 0-46-0	8.5 pounds 21-0-0
38.75 pounds 0-0-62	21.5 pounds 0-46-0
8.5 pounds 0-0-0	38.75 pounds 0-0-62

The 8.5 pounds of 0-0-0 in the first formula is a filler used to make a full measure, usually dolomitic limestone.

So what you get in the bag is not all plant food. It is a mixture of chemical elements, in the form of one or more compounds. The label guarantees a certain amount of a few things, mainly the "big three," N-P-K, but what the label *doesn't* say is often more important than what it does say. What else is in the bag? Is it toxic to crops or soil life? Does it harm the soil structure or humus? Is it available when the plant needs it? What actual fertilizer materials were used in blending the formula (if it is a blended fertilizer)? You need to know, because many, if not most commercially sold fertilizers contain harmful materials. The most often used one is muriate of potash, which is 47.6% chlorine. The fertilizer dealer should tell you (if you ask) whether his blends contain chlorine or dolomite. If he doesn't want to say, beware. State fertilizer inspectors are usually more than willing to come to your farm and take a sample of any fertilizer for analysis. If your fertilizer dealer says he is supplying a non-chlorinated fertilizer, this is how you can be sure.

Your best buy. Is the fertilizer with the highest number on the label really the best buy? The best buy in fertilizers should be the one that gives your crops the best nutrition—not just the most N, P, or K at the lowest cost per acre, but the *right amount* of nutrients, at the *right time* and in the *right forms*—and with *no harmful side effects*. You need to look at the long-term results, not just the immediate economics.

Let's compare two sources of phosphorus—a vitally important plant nutrient. Superphosphate (0-20-0) contains about 8.7% phosphorus. The label guarantees that the phosphate is in water soluble form in the bag. But the label doesn't say that nearly all the phosphate is converted to insoluble form in the soil within hours or a few days. On the other hand, rock phosphate has very little soluble phosphate, perhaps only 2%, so a fertilizer label that reads 0-2-0 doesn't look very impressive, and many farmers wouldn't buy it. It costs only about half as much as superphosphate. But it contains more elemental phosphorus, from 8–17%. Superphosphate is made from rock phosphate. It is treated with sulfuric acid to make the phosphate more soluble (resulting in a larger number on the label and a higher price—and profit to the dealer), but in a few hours, it is nearly all converted back into insoluble rock phosphate form, and your extra money was wasted.

Phosphorus becomes available to plants from the soil's large reservoir of phosphate, slowly as soil bacteria and plant roots produce various acids. Mycorrhiza fungi help feed phosphorus to roots, and animal manures make phosphate more readily available to plants. Tests done at the Virginia Agricultural Experiment Station showed that one ton of rock phosphate applied once every six years gives yields equal to superphosphate applied every year at 100–280 pounds per acre. The fertilizer costs alone per acre of the two choices (1 ton every 6 years vs. 200 pounds every year) would be nearly equal, but the extra fuel and labor costs of yearly spreading would tip the balance in favor of rock phosphate. The importance of a healthy, "living" soil is very great in helping to break down the soil's rock phosphate into available form.

In the same way as we have done for phosphorus, we could show that the best buy in nitrogen is not the form with the highest label rating (ammonia with 82% nitrogen) and the best buy in potassium is not the cheapest (muriate of potash), because these fertilizers harm the soil or soil life.

Chapter 12

LIMING & pH — SWEET OR SOUR?

W hy do farmers lime their soil? A loud chorus answers, "To correct acidity," or "To sweeten the soil." Almost everyone thinks of the purpose of liming materials as being to adjust soil pH, not to fertilize. In fact, materials used in pH and salinity control are called soil amendments, not fertilizers. There's more to it than that.

To measure acidity and alkalinity, we use the pH scale, a numerical scale running from 0 (most acid) to 14 (most alkaline), with 7 as neutral being in the middle. A change along the scale of one whole number (from 6 to 5, or from 7.5 to 8.5) means a change in acidity or alkalinity of 10 times; thus a pH of 4 would be 100 times more acid than a pH of 6.

The pH Scale

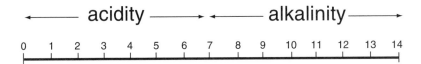

Cause of acidity. Soil acidity is caused when a large percentage of the positively charged ions (cations) held on soil colloid particles are hydrogen (H^+). Actual soil acidity that affects plants and which is measured by pH tests is caused by unattached hydrogen ions in the soil water solution, called *active* acidity. At very low soil pHs (5.0 or below), some acidity can also be due to aluminum ions, but this is not measured by a pH test. If hydrogen ions are attached to soil colloids, that is called *potential acidity*, and it is not usually measured by pH tests.

The hydrogen ions in effect displace many of the more valuable nutrient cations: calcium (Ca^{++}), magnesium (Mg^{++}) and potassium (K^+). These cations go into the soil's water and if not soon absorbed by plant roots, can leach into the groundwater. So, a very acid soil is usually an infertile soil. Higher rainfall areas tend to have acid soils, and dry climates, such as in the western U.S., usually have alkaline soils.

As plants absorb the nutrient cations, their roots release hydrogen ions in exchange (the process is called *base exchange* or *cation exchange*), which contributes to soil acidity. Measurements have found pHs over 1 pH unit lower near the root than in surrounding soil.

Testing for active soil acidity is fairly simple, and you can buy soil test kits that measure pH, or a pH meter. Soil testing labs use more exact methods, either a pH meter or special chemical buffer solutions.

Soil pH typically ranges from 4 to 10, but most crops do best under slightly acid conditions (6.0–6.8), although most crops can do well under a fairly wide range of pH, provided they receive an adequate balance of nutrients. The pH does affect the availability of nutrient elements, and at very acid or alkaline conditions, nutrient deficiencies and toxicities can occur (see accompanying graph). Aluminum and manganese are more soluble at very low pH and can be toxic. Soil pH also plays an important role in determining the ability of beneficial soil organisms to survive, for while fungi can tolerate low pH (below 6), most bacteria cannot. Thus, ordinary humus-forming bacteria, as well as nitrogen-fixing bacteria can be hampered.

So, "correcting" soil acidity can be necessary and important if the soil is very acid, but there are other things to consider. Many people do not realize that the pH can change greatly during the year. In an actual example of soil from a farm in southeastern Minnesota, the pH was 6.4 on May 28, 1982, and 4.8 on July 19, 1982. The pH can even change day to day.

Soil pH can change because of rains, because of microbial and root activity, and because of using one or another fertilizer. For example, anhydrous ammonia causes a strong alkaline pH at first, then later becomes acid, while superphosphate or high-salt-index fertilizers cause an acid pH. These pH changes can be detrimental to crops and soil life.

To avoid problems associated with very low (acid) or high (alkaline) pH soils, soil scientists recommend modifying the soil more toward the "ideal" range of pH 6.0–6.8 (for most crops) by adding so-called *liming*

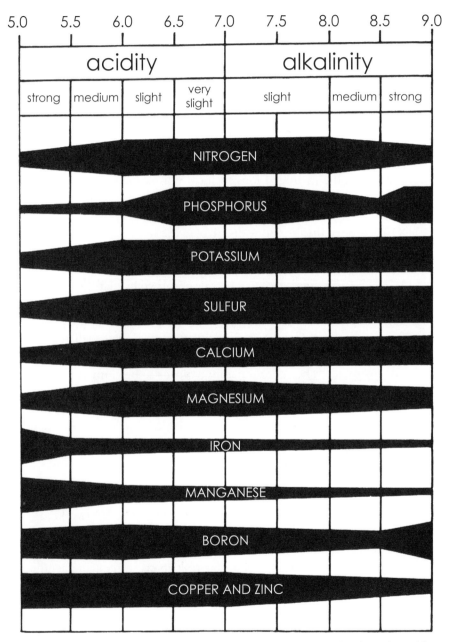

Availability of elements to plants at different pH levels for typical soils (indicated by width of bands).

materials to acid soils or *acidifying materials* to alkaline soils. (Acid tropical and subtropical soils, high in aluminum and iron, should not be limed to a pH over 5.5 to 6.0.)

Acidifying soil amendments for alkaline soils include elemental sulfur, sulfuric acid, iron sulfate, aluminum sulfate and acidic organic matter (such as peat moss, pine needles, sawdust and tree leaves). When they react with water or are decomposed by microorganisms, hydrogen ions are released, eliminating alkalinity.

Neutralization. Liming materials for acid soils include compounds that chemically are oxides, hydroxides and carbonates. When applied to the soil they react with the hydrogen ions to form non-acidic compounds, thus *neutralizing* the acidity. For example, using the common liming material, calcitic limestone (high-cal lime), which is mostly calcium carbonate ($CaCO_3$), a little of the limestone dissolves in the soil water, breaking apart into two ions, Ca^{++} and CO_3^{--} The latter carbonate ion combines with two hydrogen ions to form water and carbon dioxide: $CO_3^{--} + 2H^+ \rightarrow H_2O + CO_2$. The calcium ion is free to either be absorbed by plants or to attach to soil particles and replenish the soil's storehouse of nutrients.

Many people think that they apply "calcium" to their soil to correct acidity, but actually it's not the calcium that neutralizes soil acidity, it is the other ion that was originally connected to the calcium, either a carbonate, oxide or hydroxide. The cation calcium happens to be along for the ride. In fact, some non-calcium compounds can neutralize acidity, such as the magnesium carbonate ($MgCO_3$) that makes up part of dolomitic limestone (dolomite). Also, some calcium-containing materials do not neutralize acidity, namely gypsum (calcium sulfate, $CaSO_4$). Still, most liming materials do contain calcium.

The accompanying table of liming materials and calcium sources reveals some that are more desirable than others. Most often, ground limestone from local quarries is used, at a very low cost. But there are two kinds of limestone, calcitic and dolomitic, and calcitic is preferred unless your soil is deficient in magnesium. Excessive magnesium (contained in dolomite) can be detrimental by (1) causing some clay soils to crust, reducing aeration, (2) releasing soil nitrogen by causing formation of gaseous nitrogen oxides, (3) causing both phosphorus and potassium deficiencies in the soil, (4) causing effects similar to magnesium deficiency, (5) combining with aluminum to form a substance in plants toxic to livestock, (6) along with low calcium, allowing organic matter to form alcohol and formaldehyde when it decays, killing soil bacteria, (7) interfering with plants' absorption

Liming Materials & Calcium Sources
(compiled from several sources)

Material	Chemical Formula	% Calcium	Form	Effect on pH	Advantages, Disadvantages
calcitic limestone (calcite, calcium carbonate, high calcium lime)	$CaCO_3$	36-38	dry, liquid suspension	compare with others, below	Low price, rapidly available to plants, high calcium & low magnesium
dolomitic limestone (dolomite, magnesium limestone, aglime)	$CaCO_3 \cdot MgCO_3$	22	dry, liquid suspension	slightly greater neutralizing ability than calcite	Low price. Slowly available; with 12-14% magnesium, which can harm soil & crops.
marl	$CaCO_3$ + clay impurities	variable	dry	10-30% less neutralizing ability than calcite	Low price, little magnesium. Clay impurities.
hydrated lime (slaked lime, calcium hydroxide)	$Ca(OH)_2$	54	dry	1/3 greater neutralizing ability than calcite	Greater neutralizing effect, rapid action, readily available, little magnesium. Easily leached, high price, can cause release of nitrogen & rapid pH change.
lime (unslaked lime, quick lime, calcium oxide)	CaO	71	dry	80% greater neutralizing ability than calcite	Very great neutralizing effect, very rapid action, little magnesium. Can "burn" crops & kill soil life, high price.
calcium sulfate (gypsum)	$CaSO_4 \cdot 2H_2O$	23	dry	no effect on pH	Also supplies sulfur, useful on alkaline soils by getting rid of sodium, little magnesium.
calcium nitrate (nitrate of lime)	$Ca(NO_3)_2$	19.5-21	dry	1/4 the neutralizing ability of calcite	Also supplies nitrogen, readily available, little magnesium. High Price.
slags, basic slag	contain $CaSiO_3 \pm P_2O_5$	about 32	dry	10-40% less neutralizing ability than calcite	Low price, also may supply phophorus, little magnesium.
superphosphate (monocalcium phosphate)	$Ca(H_2PO_4)_2$ + $CaSO_4$	18-21	dry	lowers pH	Also supplies phosphorus and sulfur.
triple superphosphate (monocalcium phosphate)	$Ca(H_2PO_4)_2$	12-14	dry, slurry	neutral	Also supplies phosphorus.
rock phosphates (hard & soft)	$3Ca_3(PO)_2 \cdot CaF_2$ $3Ca_3(PO_4)_2 \cdot Ca(OH)_2$	33	dry	slightly alkaline	Also supplies phosphorus.

of calcium and potassium, and (8) by taking the place of calcium in plant cells, giving rise to poor quality crops. Also, dolomitic limestone breaks down in the soil very slowly. It may take as long as 18 months before any of its calcium enters plants. Studies on potatoes and sweet corn found that dolomite is a poor source of crop calcium. Research has found that surface-applied dolomitic lime takes 10 to 13 *years* to neutralize acid two inches deep. If calcitic limestone is not available locally, it will still be worth the extra freight expense to use it if your soil has plenty of magnesium.

The effectiveness of ground limestone depends greatly on its particle size. The finer it is ground, the more rapidly it acts, and the less that is required per acre, although fine lime is hard to spread on a windy day. The use of coarser particles has the advantage of giving a long-term activity. A mixture of coarse and fine particles would combine the advantages of both. Some people use "liquid lime," which is a water suspension of very fine particles of any liming material, able to be applied with spraying equipment.

Calcium. By far the greatest benefit of using lime is for its fertilizer value in supplying calcium, not just its pH control. Ideally for typical agricultural soils, calcium should occupy 75–85% of the negatively-charged sites on soil particles, while magnesium should fill 12–19% and potassium 1–5% (these are measured by a soil test for % base saturation). However, if one of those three nutrients is present in excess, the other two will be decreased, and possibly insufficient for normal crop growth.

A plant's need for calcium has been underestimated by many people. Many experts and books say that calcium deficiencies in soils are rare or non-existent. They base that statement on soil tests which detect all the calcium in the mineral component of the soil. But only a small part of that is readily available to a plant at one time (it slowly becomes available over the years), and many crops are deficient in calcium.

Dr. William Albrecht championed the use of lime for its calcium content. His research showed that calcium is essential for protein production in plants, it makes phosphorus more available to plants, and it is vital for animal and human health in the skeleton and for milk production. He pointed out that World War II draftees from states with naturally high calcium soils had healthier teeth and lower rejection rates than those from states with low calcium soils.

Calcium stimulates growth of soil microorganisms, including nitrogen-fixing bacteria. It is needed in the cell walls of plants and membranes of all living cells, and is required for proper nerve and muscle activity and blood clotting. Adequate calcium in the soil also can improve soil structure, by flocculating or clumping soil particles. The importance of calcium in the soil, in plants, and in animals and humans is very great and a vital part of any soil fertility program. Still another benefit of using lime is that the carbonate ion (CO_3^{--}) can eventually break down and release carbon dioxide (CO_2), which is necessary in photosynthesis.

Calcium and Disease

According to the textbook, *Soils* by R.L. Donahue, R.W. Miller and J.C. Shickluna (1983), calcium deficiencies in plants may be much more common than previously believed, including blossom-end rot in tomatoes and peppers, and tipburn in cabbage, lettuce and cauliflower. The reason appears to be a temporary deficiency of *mobile* calcium in the rapidly growing parts of plants (leaf tips, fruit). Nutrient tests of plants commonly measure *total* calcium in the whole plant, which may be adequate.

• From article by R.R. Muse & H.B. Couch, in *Phytopathology*, Vol. 65, 1965, p. 507–510:

Additional calcium decreased the disease, *Corticium* red thread, in turf grass (creeping red fescue), while high magnesium increased it. ". . . calcium-deficient regimes produced plants with significantly heightened disease susceptibility regardless of the imbalance of other nutrient elements, soil temperature, air temperature, time of plant growth, or host variety."

• From article by S. Shannon, J.J. Natti, & J.D. Atkin, in *Proceedings of the American Society for Horticultural Science*, Vol. 90, 1967, p. 180–189:

The calcium content of the seed of snapbeans is inversely related to the incidence of hypocotyl necrosis, a disorder of seedlings in which tissue of the developing root breaks down. High levels of potassium increase severity of blossom-end rot of tomatoes, blackheart of celery, tipburn of cabbage, and cavity spot of carrots.

• From University of Wisconsin-Madison extension paper 81–PF, "Calcium and Magnesium in Potato Production in Wisconsin," by P.E. Fixen and J.L. Walworth, p. 5:

". . . as tuber Ca decreases, susceptibility of bacterial soft rot in storage appears to increase."

Chapter 13

WASTE OR RESOURCE?

T here isn't much about manure to recommend it. It's sloppy, filthy, and it stinks. Most people would want to have as little to do with it as possible. But the wise farmer knows that this so-called waste material, and many other organic materials, is a potentially valuable resource which, if properly utilized, can greatly increase crop yields and quality.

Organic fertilizers. There are many types of material originating from living organisms that can be used as fertilizers and soil amendments. They include animal manures, green manures (fresh plants plowed under), seaweed and kelp, peat moss, tree bark, compost, leaf litter and leaf mold, grass clippings, sawdust and wood chips, brewery and cannery wastes, molasses residue, soybean and cottonseed meals, linseed and peanut meals, hulls and shells (of oats, rice, buckwheat, cocoa beans, and cottonseed), hay and straw, tobacco stems, castor pomace, papermill and sewage sludges, slaughterhouse wastes (bloodmeal, bone meal, hoof and horn meal, and tankage), fish meal, feathers, and humates (ancient organic deposits of salts of humic acids, somewhat like "petrified compost").

All of these organic materials are composed of mixtures of complex organic chemicals: carbohydrates, fats, proteins, amino acids, nucleic acids, vitamins, enzymes, aldehydes, alcohols, terpenes, organic acids, phenols, glucosides, esters, ketones, flavonoids, steroids, alkaloids, porphyrins, and so on—a veritable "soup" of thousands of different substances. These organic molecules are composed of atoms of the plant nutrients already familiar to us: carbon, hydrogen, oxygen, nitrogen, phosphorus, potassium, sulfur, magnesium, calcium, iron, etc.

Decomposition. When these complex molecules eventually break down into their simpler components, nutrients in the form plants can use are made available to the plants. In the soil, organic matter is broken down (decomposed) by the soil organisms, primarily bacteria, actinomycetes, and fungi. You may want to review the details of decomposition in Chapter 6.

Humus. When organic matter decomposes under aerobic conditions (with air or oxygen), humus is produced. As we have mentioned in Chapters 3 and 6, humus is an extremely valuable substance in the soil since it stores plant nutrients and makes them available to the plant slowly, as the plant needs them. It also improves soil structure, holds water, prevents erosion, and helps counteract and protect plants from excess salts, toxic chemicals, and excessive changes in pH. One of the constituents of humus, the humic acids, has been found to greatly stimulate plant root growth. The commercially available form, humates, is rich in humic acids.

The microorganisms that live in humus and help produce it, help in the above functions plus adding others: nitrogen fixation, producing a "glue" to hold soil particles together, producing plant growth hormones, preventing nutrients from leaching by temporarily making them part of their cells, protecting plants from diseases and pests, and breaking down (detoxifying) many man-made toxic chemicals.

When fresh organic matter decomposes, the "pre-humus" early material (called the *active fraction* of soil organic matter) is high in fulvic acids, polysaccharides and readily available plant nutrients (especially nitrogen). It also contributes most to soil structure.

As decomposition nears its end, true humus (called the *passive fraction* of soil organic matter) is formed. It is high in humic acids, contributes to the soil's colloidal content and is important in holding many nutrients from other sources (potassium, calcium, magnesium).

The active fraction does not last long, so more new organic matter must be added to replenish this valuable source of nutrients. On the other hand the passive fraction is very stable, lasting hundreds of years.

Compost. The value of letting organic matter decompose before putting it in the soil was mentioned in Chapter 6. If you want to take the extra time and trouble it requires, composting can greatly benefit soil and crops. There are various methods of composting, and the books and articles on the subject will generally give very specific instructions to follow. Actually, satisfactory results can be obtained without being so fussy. Just remember a few general principles:

1. What you are trying to do is provide a home for a "volunteer army" of microorganisms. They need (a) food, (b) moisture, (c) air, (d) proper temperature, and (e) absence of toxic substances.

2. The fresh organic matter serves as food. Although, to be a good food source, it should contain nitrogen (mainly proteins) and considerable carbon (mainly carbohydrates: cellulose, hemicellulose, starches, sugars). A good carbon-to-nitrogen ratio is about 30:1. Materials with high cellulose content are slow to break down (such as straw, sawdust, and wood chips), so other materials should be mixed with them. Ideally, your raw materials should include both plant and animal materials, since plant matter is high in carbon (carbohydrates) and animal matter contains more nitrogen. At least 10-20% of the compost pile, if possible, should be animal manure, which supplies bacteria, enzymes and growth promoters, and nitrogen. Low-nitrogen material such as sawdust or wood chips should make up less than 1/3 of the pile. If not enough nitrogen is available, a suitable amount of nitrogen fertilizer can be added (about 15 lbs. actual nitrogen per ton), such as fish meal, bone meal, blood meal, cottonseed meal, soybean meal, urea, or ammonium sulfate (the latter causes acid conditions; add an equal amount of lime). If you are composting mainly animal manures, slaughterhouse wastes or fish meal (high nitrogen), a considerable amount of plant matter should be added to supply carbon, absorb moisture, and provide aeration. Plant matter should not be too coarse or the pile will dry out too easily. The recommended ratio in the famous Indore method of composting is one part animal manure, sewage sludge, garbage, or green manure to two parts cornstalks, straw, sawdust, or woodchips.

3. Composting is best done in shallow (two feet) pits or on concrete slabs, but any nearly level piece of ground will do. It is best to have well-packed clay to reduce seepage of liquids into the soil. It is most convenient to either locate the compost piles near the source of materials (manure) and a water supply (if you need to dampen the piles), or else near the fields you intend to use the compost on. If you live in a rainy climate, try to orient the windrows so they do not dam up rainwater. Also, cover the pile with fibrous materials such as straw or leaves to help shed rainwater.

The organic matter should be piled or placed in such a way that a fairly large surface area is exposed. For agricultural applications, long windrows work well, about 4-5 feet high and 9-12 feet wide at the bottom, tapering to a peak or mound at top. Depending on your farm operation, you may build a windrow quickly if a lot of material is available all at once, or slowly, such as by daily cleaning of barn manure. If the windrow is built gradually, build it to full height at one end and add to it. A small front-end loader or tractor loader can be used, or some farmers throw manure out of a manure spreader. Before starting a windrow, it is good to lay down a 1 to 1 1/2 foot layer of absorbent litter, such as straw, corn stalks or cobs, sawdust or wood chips, etc., to absorb nutrient-rich liquids that seep out of the pile, and to aid in ventilation. When making the pile, be sure to mix the various ingredients as thoroughly as possible. For large-scale operations, windrows can be built over perforated pipes or hoses connected to an air blower. This allows for ventilation and for building a larger pile.

4. The material should be moist inside the pile, but not waterlogged. A moisture content of from 40-60% is ideal, but composting will occur at from 25-75% moisture. Add absorbent material (straw, ground corn cobs, shredded newspaper, etc.) to a too-wet pile. If it is too dry, add water.

5. The pile or window should be turned (mixed up, churned) one to several times during the first eight weeks or so of composting to help aerate it, mix it, or cool it. Sometimes this may necessitate turning every few days. The more often it is turned, the more rapidly composting occurs, unless the temperature in the pile falls too low (below 85°F). There are fairly expensive compost turning machines on the market, and for large composting operations, they are necessary. If you have a talent for building machinery, you can make your own. For smaller operations, a compost windrow can be turned with a front-end loader or else loaded onto a manure spreader and repiled beside the old windrow.

The idea in turning is to mix up the material and if possible put what used to be on the inside on the outside, and vice versa. This insures that everything gets composted at the same speed and that weed seeds and pathogens get killed. Don't worry about saving the bottom layer of litter (see section 3). If the material begins to dry out after turning, add water (spray the outside of the pile with a hose).

6. When other conditions are favorable, the temperature inside the pile will rise rapidly. If a pile refuses to heat up, it may be too wet or too dry, or poorly aerated. In the winter, covering the pile with a thick layer of straw, etc. will help insulate it. Temperatures from about 100°-150°F are ideal (stick a thermometer into a hole poked in the pile). If the material gets hotter than 160°F, it should be turned to cool it off, or you can poke holes into the center of the pile with a pipe or post.

7. Adding about 10-20% soil will help supply soil microorganisms to the pile, help absorb ammonia (conserve nitrogen), buffer pH and speed composting (sand and heavy clays are not good). Adding a little lime or

Windrows of composting cattle manure.

rock phosphate (2 lbs. per cubic yard) as a calcium source will stimulate microorganismic growth and prevent very acid conditions which kill bacteria. However, the pH should not go much above 7, for that will also kill microbes. If you can smell a strong ammonia odor, the material is too alkaline. Other good additives include powdered kelp (seaweed) or humates.

8. Adding a special microbial inoculant or "starter" is not usually necessary if the organic matter is already well supplied (as in animal manures), but it will speed up composting in other types of organic matter or in cold weather, so an inoculant may be worth the expense. Other good starters are some previously made compost, and compost "tea" made by soaking compost or rich soil in a container of water for several days. Any starter should be mixed or sprayed so as to make good contact with as much of the material to be composted as possible.

9. Depending on weather (mostly temperature) and the other factors mentioned, it will take from 3-12 months for compost to be formed by using windrows. It takes longer in cooler climates than warmer. Special methods can produce compost more rapidly, in 4-6 weeks. Compost is "done" when the temperature inside the pile falls to near the outside temperature and no ammonia odor can be detected.

As composting proceeds, the volume of the material will shrink considerably. Depending on original materials, from 20 to 60% of the original volume will disappear as it is "eaten" by microbes. Up to 50% of the weight can be lost.

The final result, compost, should be dark brown, crumbly, moist, rich-smelling humus, without much recognizable structure (no large stalks, stems, or other remains of the original organic matter). It should be full of beneficial microorganisms, waiting to go to work when added to your soil. It should have an abundance of plant nutrients, from major elements to trace elements to hormones and growth promoters. The percent of nitrogen will be greater than in the original organic matter (the carbon-to-nitrogen ratio goes from about 30:1 to 12:1 or 10:1), and if nitrogen-fixing microbes were active, it can be a goldmine of free nitrogen. I have found composted manure to have two or three times more readily available nitrogen than the original manure.

If finished compost must be stored for a long time, it is best to cover it to prevent leaching of nutrients or drying out. Cover with wet burlap, tarps, or plastic sheets. Finished compost should be turned and watered occasionally to prevent undesirable anaerobic decomposition.

10. Compost is essentially humus and can be put on soil at any time without harming soil or crops (however, do not use much just before a crop that does not require much nitrogen, such as small grains). It is like instant plant food. It is best to till it shallowly into the soil to keep its bacterial population active. To obtain maximum benefit from compost, do not let it dry out, but spread and quickly incorporate it into the soil (within hours). Spreading on a cloudy or rainy day is good. A convenient time to apply compost is after removal of one crop and before tillage for the next crop.

Good results have been obtained at application rates from 1/2 to 10 tons per acre. It partly depends on the nutrient needs of the next crop. If you are also trying to increase soil humus content, improve tilth (soil structure), or fight dry climate or drought, then use higher application rates. After soil is improved, much lower rates can be used.

An imaginative way of applying compost has been used by some southwestern U.S. farmers who use ditch irrigation. A pit is dug in a main feeder irrigation ditch, filled with compost, and flooded. The flow of water carries plant nutrients and beneficial microorganisms to plants. The pit should be about four feet deep (not over six feet) and can hold about 150 tons of compost. Supplementary side dressing with compost can also be done. Yields and crop quality on irrigated desert land can be tremendously improved.

Direct soil application. The other approach to using organic fertilizers is to put them directly on or in the soil rather than composting them first. This has the advantages of requiring less time and trouble, and in conserving some nutrients which escape into the air during composting (nitrogen and carbon dioxide), but the disadvantage or danger of possibly damaging a crop if done in the wrong way.

Since decomposition of fresh organic matter temporarily ties up nutrients in the cells of microorganisms, large amounts of fresh organic matter should not be applied close to planting, or plowed or injected deeply into the soil (below 5-7 inches) because it may decompose anaerobically (putrefy) and produce toxic substances. Large amounts of animal manures can cause salt damage and an over-supply of potassium if the manure cannot decompose aerobically.

Manure management. For those farms with livestock or poultry, manure is a natural by-product, a "free" source of organic matter which should be used as a valuable source of nutrients. Unfortunately, many farmers do not make full use of its potential, or mis-use it with harmful results, and some even make no use of it at all.

Like other organic matter, fresh manure contains nitrogen, phosphorus, potassium, and other elements that *could* be used by crops if the manure is handled and applied in ways that conserve its nutrient value. But by using wrong management methods, up to 80% of the nitrogen can be lost—or even nearly all of the nutrients—or else toxic salts and other toxic decomposition products can damage crops. Again, by using certain common-sense principles and by understanding soil ecology, you can avoid the pitfalls and reap the benefits of using manure.

There is a variety of manure handling, storage, and application methods, ranging from daily spreading to stacking of solid manure to storage of liquid manure slurry, later spraying it on or knifing it into the soil. The different manure handling systems have their own advantages and disadvantages, both with regard to required time and equipment, and the nutrient value when it finally is used by crops. The following general manure management principles can be applied to any system:

1. As with other organic matter, manure must decompose aerobically to form humus in order to best benefit crops. It should be worked into the aerobic zone of the soil (upper several inches) where conditions are favorable for aerobic microorganisms.

2. Thinner applications over more acres are more effective than larger amounts at once, which can smother plants and cause salt damage, nitrate poisoning in animals, and pollution of groundwater, or become anaerobic and produce other toxic substances.

3. Apply most manure in the fall, so it has time to decompose before crop growth and does not compete with crops for nutrients.

4. Save the liquid portion (urine), since it contains valuable nutrients.

5. The use of bedding increases the value of manure and also absorbs urine.

6. Avoid or reduce nutrient loss by:
   a. Working manure into the soil as soon as possible after spreading.
   b. Spread on cool, rainy days if possible.
   c. Do not spread on snow or frozen ground, especially on sloping ground, to avoid nutrient run-off into streams.
   d. Do not stack manure where leaching or loss of urine can occur. A concrete floor is helpful.
   e. Additives can be used to tie up ammonia, such as superphosphate (0-20-0) or humates; they should be added to manure as soon as possible and mixed as thoroughly as possible.

Up to Date?

From *The Farmer's Every-day Book*, by J.L. Blake, 1851:

How to Preserve Manure—"If you cannot use all your manure, place it in heaps and cover it with earth two feet in thickness, which will inhale and retain most of its enriching gases till wanted. Nevertheless, it is better every year to put on your land all the manure you possibly can scrape together. Leave not a particle in your barnyard. All you can get from it before another season is clear gain, for it will lose but little more underground with a crop over it, than exposed to the action of the sun and atmosphere."

Making Manure—"A correspondent of the Maine Farmer says that he converted straw, corn-stalks, and potato-tops into good manure in fourteen days, only by heaping them together, and mixing unslaked lime with them. He used six casks of lime, and had fifty loads of good manure. The loads were such as farmers usually carry, a large half cord to the load."

How to Get Manure Cheap—"As soon as the summer manure is scraped out of the cow-yard, spread over it again plentifully, leaves, litter, loam, peat-mud, or almost anything else. Fail not to neglect it. The cost to do it will be much less than to purchase guano or poudrette, and it will be more useful, for you will be able to improve the texture of your soil, if you exercise good judgment in selecting your materials."

    f. Composting, if possible, will temporarily tie up nutrients in microbial cells and humus and keep them from being lost.

  7. If additional materials are added to increase manure's fertilizer value (green manures, fertilizer supplements, lime, rock phosphate), mix them in as thoroughly as possible. Tests have shown that such supplements greatly increase the nutrient value of manure, more so than applying the materials separately.

8. "Hot" manures (poultry, pig, horse) should be used with care to avoid plant injury. They can be mixed with cow manure, composted, or applied in small amounts well before crop growth.

9. Rates of fresh manure application depend on soil conditions, crop to be grown, time of year, and kind of manure. Up to 10 tons per acre of solid manure is generally safe (2500 gallons per acre of liquid slurry), but less may give better results. "Hot" manures should be applied at lower rates.

Organic vs. synthetic. Organic fertilizers have a considerable number of advantages compared to synthetic "salt" fertilizers (see Chapters 10-11). Of course, some salt fertilizers are worse than others, and if properly used, the safer ones can give good results. But in general, the use of synthetic fertilizers alone can lead to a number of problems:

1. Can be toxic to plants and soil life (salt damage, chloride), or animals that eat the crops (nitrate poisoning).

2. Can destroy humus (excess ammonia).

3. Can destroy soil structure (tilth) by destroying humus or causing crusting or hardpan.

4. Can pollute surface and groundwater (nitrates, phosphates).

5. Can upset nutrient and pH balance in soil; they provide only a few nutrients; plants may be unhealthy and vulnerable to diseases and pests; crops may be of low quality.

6. Provide only a quick "shot in the arm" of some nutrients (not long, slow release).

If organic fertilizers are used improperly, they also can cause problems, such as pollution and salt damage from animal manures, but on the whole, organic fertilizers and natural inorganic minerals offer the following advantages:

1. Are safe for plants and soil life; stimulate soil life.

2. Build up humus.

3. Improve soil tilth.

4. Hold nutrients, do not pollute water.

5. Feed plants a "balanced diet"; often provide trace elements and growth-promoting substances; produce healthy plants that are usually resistant to diseases and pests, and high quality crops. Increase soil's buffering capacity to resist pH change and absorb toxic substances or detoxify them.

6. Release nutrients slowly, in greater amounts in warmest months, when plants need more nutrients.

But one should not conclude from the above lists that synthetic fertilizers are totally bad and that organics and natural inorganics are totally good. Anything good can be mis-used. Some synthetics are definitely harmful and should be avoided (see Chapter 11), but others have their place. However, synthetic salt fertilizers should not be used alone. Any farming operation should always include the use of some organic and natural inorganic fertilizers and soil amendments. Tests have proved that mixing rock fertilizers or synthetics with manures gives much better results than using either separately, or even together but not mixed. Experiments in Russia found that potatoes grown on sandy loam soil yielded 7,930 lbs./acre without fertilizer (12-year average); 15,325 lbs./acre with 16 tons of manure per acre; 16,483.5 lbs./acre with mineral fertilizers corresponding to the N,P,K,Ca content of 16 tons of manure per acre; and 18,265.5 lbs./ace with 8 tons of manure per acre plus mineral fertilizers corresponding to the N,P,K,Ca content of 8 tons of manure per acre. (E.M. Bodrova & Z.D. Ozolina, *Simultaneous Application of Organic and Mineral Fertilizers*, 1965).

Chapter 14

SCRATCHING THE SURFACE

P roper use of fertilizer is certainly important, but tillage methods and cropping systems—what we do to the soil, and when—play a large role in producing optimum yields. An additional major consideration *should* be: what *long term effect* does the cropping system or tillage method have on the soil and soil life? No conscientious farmer can afford to do anything that will eventually ruin the soil, even though he may get fantastic yields (for a while). Unfortunately, not all of the textbooks, agronomists, and extension agents look that far down the road. They let bushels and dollars take precedence over organic matter and soil life. In the textbook, *Soil Fertility and Fertilizers* by S.L. Tisdale and W.L. Nelson (2nd ed., 1966), we read: "Maintenance of organic matter for the sake of maintenance alone is not a practical approach to farming" (p. 564); and in a section on whether legumes should be used to supply nitrogen, or commercial fertilizers, we read: "The selection of the program that the farmer should follow becomes a matter of economics; he should choose the program yielding the greatest net return on his investment" (p. 566).

Proper fertilization and tillage methods can give terrific yields AND not harm the soil, AND produce *high quality* crops at the same time— something that "traditional" agricultural methods almost never can do. Let's look at some of the commonly used cropping systems and tillage methods. Many factors should be considered: soil type and condition (compaction, hardpan, temperature, herbicide carryover, soil erosion, slope and drainage, subsoil), crops to be grown and their nutrient requirements, root depth, organic matter returned to the soil (plant residues, manures), soil fertility and nutrient balance (soil test results), weeds and pests, planting depth and time, plant population and row width, tillage and seedbed preparation, and markets.

Is rotation necessary? For decades farmers have been advised to rotate their crops to prevent depleting their soil, and now we sometimes hear the "experts" tell us not to rotate. Which is best?

There are advantages and disadvantages to either system, and it depends on your local situation and needs. First of all, if the soil is properly fertilized and if organic matter and soil life are fostered, it is possible not to rotate. Each crop species has its own set of nutrient requirements and soil type preferences, and if you get your soil "fine tuned" for one crop, and if it pays to keep raising it (if market prices hold up, or if you need it to feed livestock), you can raise the same crop year after year. As we will see in Chapter 16, even weeds and insects do not have to be a problem in properly fertilized soil.

Rotations usually include a legume crop to add nitrogen to the soil (especially alfalfa and clovers, but not soybeans) and a variety of deep and shallow-rooted crops with the intention of improving tilth and utilizing nutrients from different depths. These are good reasons, and they do have their value; but, if you don't want to or need to, you don't have to rotate. A possible exception is reported in *Soil Conditions and Plant Growth*, 9th ed., by E.W. Russell (1961): continuous clover causes an accumulation of a toxic substance secreted by clover roots which will inhibit plant growth; however, it can be counteracted by adding large amounts of manure.

In tests done at the University of Missouri Agricultural Experiment Station, growing a corn-wheat-clover rotation from 1917–1941 on soil which had previously been cropped for 60 years did not increase the amount of nitrogen significantly (0.136% in 1917 vs. 0.137% in 1941), while a continuous bluegrass sod increased nitrogen from 0.105% in 1917 to 0.132% in 1941, and red clover turned under increased nitrogen from 0.137% in 1917 to 0.170% in 1941. The latter two crops would allow humus to build up in the soil, while the rotation did not. Some rotations are definitely harmful. According to University of Minnesota soil scientist G.R. Blake, a series of row crops (which require a lot of tillage) can deplete the soil, since not enough plant residues are returned and too much tillage destroys soil structure (*Crops & Soils Magazine*, March 1980, p. 10–11).

Continuous cropping is said to encourage the build up of soil-harbored diseases, weeds and pests. Yet continuous corn has been grown at the University of Illinois test plots since 1876 with no indication of an increase of insects or diseases. Continuous cropping can destroy soil tilth and encourage erosion (if little organic matter is recycled or if soil organisms are destroyed), can cause more soil compaction from increased cultivation and harvesting traffic, and can deplete organic matter from increased cultivation. These dangers can be minimized or overcome by a "living" soil with plenty of humus. With plenty of humus and good soil tilth, and by use of certain tillage methods to break up a hardpan, roots will grow deeper and utilize subsoil nutrients and be better able to withstand drought.

Tests Run At The North Carolina Agricultural Experiment Station Supplementing Corn With Varying Amounts Of Nitrogen

Nitrogen (lbs/A)	Grain (bu/A)	Cornstalks (lbs/A)
0	22.8	3,614
40	52.1	3,356
80	81.7	5,078
120	100.5	5,459
160	102.3	5,328

Continuous cropping has the advantage of allowing forage crops to be grown on sloping land to reduce erosion and row crops to be grown on level land. These two types of crops require a different balance of soil nutrients (alfalfa requires high calcium), so the soil can be fertilized for one crop, not several. If you want to specialize on only one crop, you can, and you will need less machinery than for several crops.

On the other hand, rotation has many advantages. It does improve tilth and, if properly done, can build up soil nutrients. It uses natural crop plants to do these things rather than commercial fertilizers. It will take longer to get results. Also, tests have shown that supplementing crop residues with commercial fertilizers gives increased yields (as long as you are careful to fertilize properly to keep nutrients balanced and quality high) and increases the amount of stalks (plant residues) that can be recycled to form more humus. Many farmers have found that a shorter rotation of forage or grass crops for only a year at a time (or two for hay) in rotation with row crops works better than longer rotations. A faster change of crops often results in less weed pressure.

So whether you want to rotate crops or continuous crop is up to you. Either cropping system can have advantages, IF you use proper methods and do not deplete soil fertility, humus, and organisms.

Intercropping. There is yet a third type of cropping system, interplanting or intercropping, the practice of planting two crops together, which are compatible and work together (gardeners often use the term, companion planting). Planting a nurse crop of small grain, peas or flax along with alfalfa is a time-honored example of intercropping.

With a crop such as corn, rather wide-spaced rows are used, and another crop such as soybeans, forages, small grains or sod is planted between the rows. This practice has the advantage of excellent weed control, plus often the added yield of the second crop. If a forage or grass is to follow corn in a rotation, it will already be established. A legume will add extra nitrogen in good soil. The yield of corn is not greatly hurt (except that wider rows are used), but the growth of the other crop will be reduced by the shading of the corn. If soil moisture is low, intercropping would cause too much competition for water, however. Alternating two or four rows of corn (with narrow spacing) with four or six rows of soybeans often works well, since the soybeans are less shaded.

Interplanting a small grain (winter rye or barley) and/or a forage legume (clovers, hairy vetch, etc.) with corn will give good ground cover over the following winter. Then in the next spring, plowing in the cover crop will provide nitrogen and organic matter for the next crop. The cover crop can be interseeded during the last cultivation of the corn (about a foot high).

Intercropping different plant species or even different varieties of the same species can reduce insect pest attack, perhaps because the different odors of the two crops confuse the pests, or sometimes because some plants produce repellent odors. In one experiment, tomatoes planted along with Brussels sprouts reduced the number of eggs laid by two insect pests of Brussels sprouts.

To till or not to till? In recent years, no other farming method has been so ballyhooed as reduced tillage systems, whether called minimum-till, conservation-till, or no-till. It is promoted by farm papers and magazines, fertilizer and pesticide literature, universities and their extension agencies, the USDA, and even ordinary newspapers. Local farmers' testimonials are usually included. The advertising has been working. The percent of cropland put in conservation tillage has grown from only 2–3% in 1965 to 35% in the mid-90s to perhaps over 50% today.

The two big problems which reduced tillage is promoted to overcome are (1) soil loss by wind and water erosion, and (2) energy—fuel and labor costs. Additional advantages are increased water absorption (and decreased evaporation), being able to farm more sloping land (by reducing the erosion danger), being able to plant and harvest under a wider range of soil moisture, and needing less machinery (no plows, no large tractor, no cultivator, etc.).

Types of reduced tillage range from no-till with planting done in last year's stubble; to ridge-till, with ridges and valleys; to mulch-till, where shallow tillage leaves crop residue on a cloddy surface. Basically, reduced tillage systems leave a considerable portion of plant residues on the surface (at least 30% of surface covered) and use a minimal number of passes over the field. The plant residues act as a mulch and reduce wind and water erosion, and aid in water absorption. Water erosion can be reduced by as much as 50-fold, according to a study in Ohio. Fewer passes over the field obviously save fuel and labor. All of these advantages exist.

In a study of no-tillage agriculture by several University of Kentucky professors (reported in *Science*, June 6, 1980, p. 1108–1113), the total energy required for no-till versus conventional tillage methods, from machinery needed to fertilizers to seed to herbicides and pesticides to planting to harvesting, was calculated. They found that no-till methods would use 4.39 million Calories of energy compared to 4.72 million Calories for conventional tillage, a 7% reduction. The main saving in energy of no-till is in planting costs. A University of Nebraska study showed that conventionally tilled corn required 5.33 gallons of diesel fuel to plant, while no-till corn only needed 0.90 gallon. Sounds great, doesn't it? But let's look at the rest of the story.

There are many disadvantages of reduced tillage and many problems it *creates*. First, a farmer can't just go out and use conventional tillage machinery and methods in a reduced till fashion. Conventional planters and drills generally do not work well with plant residue covering the surface, so you have to buy a new planter. The plant residues form a protective overwintering home for insect pests and plant diseases, so more pesticides may be required (about 50% more for corn). The surface mulch reduces the soil temperature in the spring (as much as 10.8°F lower at 1 inch depth), so you need a cold-resistant variety if you live in the central or northern U.S. The lack of tillage encourages weeds, especially grassy ones,

A no-till field with old stubble on the surface.

so more herbicides are required. Fertilizers and soil amendments have to be applied on the surface, so more fertilizer must be used to allow enough of it to work its way down to the roots (some planters can place fertilizer deeply in the row). Surface application of nitrogen fertilizers causes low pH in upper soil layers. Recommendations now are to split the application of nitrogen, with some applied with the main fertilizer application at planting and the rest applied as side dressing later; this means an extra trip over the field! Emergence percentage for reduced-till field is usually lower than for well-prepared seedbeds (10–15% more seed is required for no-till). No-till crop roots are shallow, possibly a disaster in a drought. Yields in some cases are lower with reduced tillage (for corn on poorly drained soils). By keeping most of the plant residues on the surface, most of the potential nutrient value from their decomposition is lost by oxidation-mineralization (see Chapter 6), and they do not contribute much to soil tilth. No-till methods do not work well if the soil pH is below 6.6, but most crops do best from about 6.0–6.8.

With all those disadvantages, why are reduced tillage methods promoted so heavily? The necessary and increased use of herbicides and pesticides, and the higher amounts of fertilizer needed may explain why the fertilizer and chemical companies are pushing reduced tillage. If we consider the long term effects of poisoning the soil with herbicides and insecticides, killing soil life, and depleting the humus, whatever dollars are saved in the short-term by reduced tillage just may not be worth it! In some parts of the country, with certain soil types, reduced tillage has been made to work, but it's not for everyone.

A better way. Wouldn't it be great to have most of the advantages of reduced tillage and none of the disadvantages? You can—by using the natural systems approach to agriculture presented in this book, plus certain tillage methods. As we have mentioned in Chapters 3 and 6, a soil with plenty of humus and soil organisms will resist water and wind erosion (remember the "glue" secreted by the microorganisms?) and will be spongy and will soak up and hold a lot of water. Disking plant residues under, with a small amount sticking out to the surface will prevent erosion and aid water absorption by a wicking action, plus the below-ground organic matter will decompose into nutrient-rich humus. Soil with good tilth will be easier to work, resulting in some fuel savings and obviating the need for large tractors. Humus-rich "living" soil will fight soil insect and nematode pests and plant diseases. Soil rich in organic matter warms up faster in the spring. Ridge planting, where the soil is formed into a ridge every 40 inches, is an effective tillage method which helps conserve moisture and helps warm up the seedbed on the tops of the ridges (in drought-prone areas, the seeds should be planted between the ridges in the "valleys").

It is wise to take the temperature of your soil in the spring, for some seeds (corn and soybeans) will not germinate until the temperature at planting depth reaches about 55°F, or better yet 60°. Take the temperature in the morning, because direct sunlight can raise the temperature 8 or 10°F.

In row crops, high plant populations and narrow rows may or may not pay off. More plant nutrients will be needed, so the soil must be at peak fertility. Some hybrid varieties produce higher yields in higher populations and some do not. In dry years, high plant populations release considerable soil moisture by transpiration through the leaves.

Hardpan. To farm more of your soil, break up a hardpan by plowing an inch or two deeper at a time or use a chisel plow or subsoiler. Plowing under lime will help improve the soil tilth and help overcome a hardpan. Even deep-rooted plants such as alfalfa cannot effectively penetrate a dense hardpan. Soils have a large store of nutrients "on reserve" in the subsoil, so why spend extra dollars for fertilizers when you could use what is already there? A hardpan also severely limits water absorption, leading to scenic-looking lakes in wet weather.

Weeds. Weeds can in part be controlled by two tillage methods (see Chapter 16 for other weed control methods). Weeds can often be controlled by making sure the crop gets out of the ground first and out-competes and shades the weeds. This can be done by preparing an optimum seedbed *in the row*, but leaving the soil between the rows in poor condition. The other method of weed control by tillage is the familiar use of cultivation during the growing season, being careful not to cut the roots of the crop plants. Cultivation has the advantages of not only killing weeds, but also of breaking a surface crust and helping aerate the soil. However, too frequent cultivation wastes fuel and tends to destroy humus by causing too rapid microbial decomposition.

Plowing. Fall plowing is preferred over spring plowing because freezing and thawing help improve tilth, plus hibernating insect pests are more likely to be killed from exposure.

For ordinary tillage and seedbed preparation, deep moldboard plowing is not recommended. It turns the soil completely over, covering the former aerobic top-most soil and possibly making it anaerobic, and it brings up relatively lifeless anaerobic soil. The whole soil "ecosystem" is disturbed, and it takes time to readjust. It is better to chisel plow or disk. Tilling too-wet soil causes clod formation and compaction from machinery. Occasional plowing is useful in good, aerobic soil since it increases the depth of aeration and works humus down to enrich the subsoil. Shallow moldboard plowing (just below the aerobic zone) in the fall is a beneficial practice. The depth of the aerobic zone can generally be determined by noting how far down the small feeder roots and root hairs are on a plant you have dug out of the ground.

Of course, on sloping land such common-sense conservation practices as contour plowing, grass waterways, and strip cropping are beneficial, although one should ask himself if planting row crops on steeply sloping land is wise land use. To reduce erosion, forage or pasture usage should be followed instead.

Multiple cropping. One cropping system which offers the promise of greatly increasing the productivity of an acre of land is multiple cropping, growing two or more crops in one growing season. Multiple cropping has long been practiced in the southern U.S., but recently has been used more in the North. It is promoted along with reduced tillage systems since as much time as possible should be saved, and plowing and seedbed preparation are eliminated in reduced-till methods. We have already covered reduced tillage; multiple cropping can be done just as well using conventional tillage methods, provided one is careful to maintain soil fertility and tilth, and provided the soil's water supply is adequate.

Depending on the latitude (climate), various crop sequences can be used: In the South, winter wheat-soybeans works well, as does wheat or barley followed by corn for silage, or small grains for silage followed by corn for grain. In some cases, even as far north as Wisconsin, it is possible to double or triple-crop by growing one or two crops of vegetables, which have a short maturity time, followed by a winter cover crop, such as rye. Often, one crop can be aerial-seeded into the preceding one while it is still growing, such as seeding wheat into standing soybeans.

A major factor in multiple cropping is whether the soil has enough fertility to supply the crops' needs. The right sequence of crops, such as a legume or green manure crop before a high-need crop, is prudent since commercial fertilizers have shot up in price. Also, too many crops in one year require irrigation in dry years. Planting a cool season crop such as wheat in the spring or fall can increase productivity per acre as much as 30%. Soil with a high humus content warms up faster in the spring. A light application (200 lbs./acre broadcast) of ammonium sulfate in the fall will do the same.

Chapter 15

IS "ORGANIC" THE MAGIC WORD?

B y now, you probably realize that we do not advocate the use of toxic agricultural chemicals that kill soil organisms and pollute the environment, and that we recommend building up the organic matter and humus content of the soil. Does that make us "organic farmers"?

What is "organic" farming? J.I. Rodale's *Encyclopedia of Organic Gardening* defines it: "Organic gardening or farming is a system whereby a fertile soil is maintained by applying Nature's own law of replenishing it—that is, the addition and preservation of humus, the use of organic matter instead of chemical fertilizers, and, of course, the making of compost and mulching."

Organic farming and gardening can be described as a movement and a philosophy, which has a growing number of "converts." A strict adherent to organic methods absolutely refuses to use any "chemical" fertilizers, or pesticides or herbicides, and will only use "natural" fertilizers, soil amendments, and pest, weed and disease control practices.

What we call organic agriculture is simply an outgrowth of the practical, low-tech methods that indigenous cultures have perfected over thousands of years. These people were in tune with their local environment and learned by generations of experience how to best raise plants and animals.

The pioneer of the modern organic movement was Sir Albert Howard, a British horticulturist who was director of the Institute of Plant Industry at Indore, India. He developed a method of composting agricultural and city "wastes" called the Indore method. This is described in his book, *An Agricultural Testament* (1943). (He first published his Indore method in an earlier book, 1931.) Since then, disciples of the organic movement have "spread the word" and have made many "converts" on all continents. And with good reason, for organic agriculture is based on ecologically sound principles. It seeks to obey and work with the laws of nature, not to modify and violate them. Many things organic agriculture followers do is good—but do they do enough?

Yes, but. . . Many organic farmers and gardeners have a basic mistrust of anything "scientific," associating science with "chemical" fertilizers and pesticides. Consequently they often do not have their soil tested and do not have a basic understanding of soil ecology, soil chemistry, and plant functions and nutrition. They assume that using "natural" fertilizers and soil amendments will automatically supply all of the plant's major, secondary and micronutrient needs. Now, completely organic methods MAY supply all of the plant's needs, but then again they may not. How can you tell unless you know what the needs are and unless you test your soil frequently enough to see what is going on? An organic farmer may happen to be doing everything just right and may have fantastic yields of high quality crops . . . BUT, some organic farmers have out-of-balance soil and poor quality, moderate-yielding crops. Some organic farmers get good results the first several years they switch to organic methods, but then their yields and quality gradually decline and pest problems increase. They are "organic by neglect." Another way to be technically organic is by using input substitution; that is, substituting organically approved fertilizers for conventional ones, but not following all of the vitally important sustainable cultural methods, such as good rotations, cover cropping and soil building.

If we review what we learned about decomposition in Chapter 6, we recall that when soil microorganisms decompose organic matter, plant nutrients are temporarily "tied up" in the microbes' cells, and the nutrients only become available to the plants *slowly*. Also, microbial activity can be slowed or stopped by the wrong temperature or pH and lack of oxygen or moisture. Natural rock fertilizers and soil amendments (lime, rock phosphate, greensand, etc.) also break down slowly. So unless conditions are ideal, a crop may not receive enough nutrients, especially during the critical maturing period. This is very common in these days of high-yielding hybrids and high plant populations. Thus, in some cases, organic matter alone can't do it all.

The relatively high market prices earned by organic farmers may convince many to go organic. If so, great. But if you don't want all the extra work, you can follow a middle path, namely, use organic methods as much as possible, but SUPPLEMENT the organic matter and humus with certain synthetic fertilizers, if necessary (using frequent soil tests to monitor conditions). But only those synthetic fertilizers that do not harm soil or soil life, and only in "safe" amounts.

Five-year tests run at the Vermont Agricultural Experiment Station by F.R. Magdoff showed that manure application to corn should provide all of the nitrogen needed by the crop, but that in finer textured soils (clays) and in wet years, not enough nitrogen can be provided by ordinary manure application rates, and supplementary inorganic nitrogen (ammonium nitrate) is effective (*Agronomy Journal*, July–August 1978, p. 629–632). Also, don't forget that whenever part of a crop is harvested, removed from the land and not recycled, some nutrients are removed from the soil and not replaced. If a deficiency develops (it may not; frequent soil tests should tell), supplementary fertilization above and beyond organic matter may be necessary.

Note the results of the North Carolina tests of supplementary nitrogen on corn yields (box in Chapter 14). Similar results are obtained by supplementing other nutrients. Of course, university tests usually are not run under strict organic/no-toxic-chemical conditions, and scientists are concerned only about yield, not quality. We will discuss how quality can be measured in Chapter. 18.

Tests Run at the North Carolina Agricultural Experiment Station Supplementing Oats with Varying Amounts of Phosphorus

Phosphorus (lbs/A)	Grain (bu/A)	Straw (lbs/A)
0	15.2	1,108
4.4	86.7	5,730
6.2	85.3	5,537
11.0	121.6	6,542
22.0	126.3	10,733
33.0	127.5	8,246

Natural vs. synthetic. Most standard textbooks, professors, extension agents, and fertilizer salesmen say that there is no difference between crops grown by organic methods and "chemical" methods—that the nutrients used by plants are the same inorganic ions, no matter what source they came from. That the food value of the crops is the same—no difference in vitamin content, proteins, etc. That vitamins derived from living organisms are identical to man-made vitamins. This argument seems true to a chemist and is reassuring to those involved in the synthetic fertilizer and vitamin business—*but* "it ain't necessarily so."

Some things the chemist doesn't measure are flavor and texture. Vegetables raised by N-P-K technology on "dead" soil are typically tough, fibrous, and tasteless, while those grown on humus-rich soil are tender and flavorful. Plants grown organically contain more vitamins than those grown with other fertilizers. A recent survey of 41 studies comparing organic vs. conventional fruits, vegetables and grains revealed that organic crops are significantly higher in vitamin C, iron, magnesium and phosphorus, and lower in nitrates. Both humans and animals prefer compost-grown food, and a smaller quantity of such food (up to 15% less) is needed to produce a given weight gain.

Dr. William Albrecht experimented at the University of Missouri by fertilizing timothy hay with extra synthetic nitrogen. The extra nitrogen increased the "crude protein" content of the hay (crude protein is found by multiplying the amount of nitrogen found by chemical analysis by 6.25). But when the hay was fed to weanling rabbits, 70% of them died during a summer heat wave. When the experiment was repeated and the rabbits were switched to red clover hay grown without extra nitrogen, no more deaths occurred. Although the two hays had nearly equal amounts of *total* amino acids, the *balance* of amino acids was different (*The Albrecht Papers*, 1975, p. 171–173). The excessive salt fertilizer apparently upset the plant's metabolism and caused it to produce a balance of amino acids which was not conducive to good health in animals.

It has been found that excessive levels of nitrates and nitrites in livestock (at subtoxic levels) cause depressed thyroid gland activity, leading to reduced conversion of carotene to vitamin A, leading to reduced ability to synthesize RNA (ribonucleic acid), which is necessary for normal cell metabolism (*Unfit for Human Consumption*, by Ruth M. Harmer, 1971). Dr. Barry Commoner of Washington University, St. Louis, has found that synthetic nitrogen compounds have a different isotope make-up than natural nitrogen compounds. Plants grown with synthetic nitrogen may fail to field ripen and may be readily attacked by insects.

Organic is Coming On

Although organic gardening and farming were snickered at and dismissed as impractical back in the 1960s and '70s, the economic, environmental and health disadvantages of conventional, high-tech agriculture have helped propel sales of organic products upward 20% or more per year, in spite of big-agriculture's best efforts to undermine it. In fact, huge conventional ag and food marketing companies have seen the handwriting on the wall and are trying to capture some of the approximately 11 billion U.S. consumer dollars spent on organics each year. They put out products with labels proclaiming, "All Natural," "USDA Organic," "BGH-Free," "No Preservatives," "Zero GMOs" and so on. Some truly are produced by ecologically sound methods, while others may be only technically organic, such as cattle or poultry being called "free-range," which only means that the poor concentration-camp-raised animals had *access* to the outdoors; that is, their keepers opened a door to the outside pen occasionally, but the indoor-accustomed critters were probably afraid to venture out.

Less poison. With average consumers more and more concerned about the quality and possible toxic contamination of their food, it is no wonder that organic products are doing well, even though organic food is usually more expensive than conventional. A study reported in *Food Additives and Contaminants*, May 2002, compared pesticide residues in organic and conventional food as tested by the USDA and Consumers Union. Over 94,000 samples from more than 20 crops were tested. From 73 to over 90% of conventionally grown fruits and vegetables had pesticide residues. Unfortunately, 23 to 27% of organic products also had residues, believed to be due to wind-blown pesticide drift or else soil contamination by now-banned persistent pesticides in the past.

Certified ... or not? One difficulty farmers have in becoming organic is the array of "hoops" they must jump through to be officially certified. The details are different in different countries, and local organic organizations and contacts can be found on the internet or from acquaintances. Typically a farm must go through a three-year transition period during which organically-approved inputs and methods are used. The transition period is intended not only to allow the farmer to work out a viable system for the farm, but to allow the soil to cleanse itself of toxic substances and begin to build itself up. There will be inspections by certifying representatives and fees or dues to the certifying organization.

During the transition period, depending on connections, a farmer may be able to get better than usual market prices, although full premium prices probably will not come until certification. Sometimes a farm can be helped through the process by being a part of an organic marketing cooperative.

Many ecologically-minded farmers believe that the process to become certified is worth it, such as when their products are being sold to companies or co-ops that require certification, but others, who may sell relatively small amounts locally, perhaps to restaurants or at farmers' markets, may want to farm organically without the trouble and expense of certification.

An excellent and rapidly expanding market is community-supported agriculture (CSA), where an organic (or sustainable) farm sells weekly boxes of whatever produce is in season to hundreds of local customers who pay an annual membership fee. Customers routinely praise the freshness and flavor of locally-grown food.

The future of organic. The extraordinary growth and popularity of organically-grown food and fiber has made the previously dismissive practitioners of conventional agriculture sit up and take notice. Any enterprise that brings in billions of dollars and expands at 20% annually commands respect. Once "orthodox" university ag colleges now have professors and research programs dedicated to sustainable systems. Organic co-ops and farmers' markets have sprung up like weeds. Books and magazines on organic subjects are readily available. Homemakers and up-scale restaurants cook with organic ingredients. Even corner supermarkets have an organic foods section.

But with all this success and growth comes uncertainty and danger. Will the present-day fad-like popularity of organic products die out? Will organic agriculture become too successful for its own good?

Now in the early 21st century, the main danger seems to be degradation of traditional small-scale production and strict organic standards. Not wanting to pass up any source of profits, big agribusiness is trying to appropriate the organic market. In 1990, the Organic Foods Production Act put organic standards and research under the USDA, with major input by a group of organic farmers and marketers, along with environmentalists, called the National Organic Standards Board (NOSB). In 1997, changes in standards not approved by the NOSB were proposed that would allow genetically engineered and irradiated materials, along with sewage sludge, to be used to grow "organic" products. The resulting public outrage was overwhelming, and the USDA quickly withdrew the changes. New nationwide organic standards were formulated in 2001. Other countries and organizations have their own standards.

But the assaults continue to come. Some of the huge food processing and marketing corporations have been trying to cut corners and still call their products "certified organic." They follow the letter of the standards, but not the spirit, as mentioned earlier in this box. What will organic agriculture become? If consumers stay vigilant and do not let unscrupulous operators corrupt it, organics will have a bright future.

Does It Pay?

All right, you say. So natural farming methods can produce better crops and healthier animals. But can I make a profit at it? Won't my yields drop if I stop using pesticides and anhydrous ammonia? Don't the special low-toxic fertilizers cost a lot more per ton?

With on-farm incomes often falling behind national averages and family farms becoming an "endangered species," and with typical big and small farms in debt up to their eyeballs, it is no wonder that an average farmer would be hesitant to radically change his whole farming system. We will say more about changing systems in later chapters, but for now, *does it pay?*

At least since the 1970s, studies have been done comparing input costs and profitability of organic or sustainable vs. various conventional systems. Overall, the results come out in favor of natural-system agriculture. Often the yields are lower on natural-system farms, especially in normal or wet weather and during the transition period out of conventional methods, but in dry years, natural-system yields usually top the conventional yields. Corn and small grain yields tend to be lower, and forages higher with natural methods. After the transition period, when soil becomes much healthier, organic and sustainable yields seldom are lower than conventional.

Then there are the inputs. The special fertilizers and soil amendments that natural systems use often cost more per ton . . . but usually much lower application rates are used per acre since a great deal of crop nutrients are supplied by compost, manure, green manures and beneficial soil organisms. Sometimes additional trips over the fields require a bit more fuel than in conventional systems, but then the natural systems use little if any expensive pesticides or herbicides, although there can be more labor needed for weed or pest control. In total, operating inputs normally cost less per acre for natural systems than conventional.

Next, the prices obtainable by natural-system farmers for their products are often higher than for conventionally-grown commodities, although in some cases selling or marketing naturally-grown products is more labor-intensive, as in a farmers' market, community-supported agriculture, or selling to local restaurants or supermarkets.

Finally, there is the broad-scale, long-term difference in farming with relatively low-energy, safer, health-promoting methods and materials, compared with capital-intensive, fossil-fuel wasting (in commercial fertilizer and pesticide manufacture), health-risky methods and products (these last factors are rarely considered in comparison studies).

But let's cite some specific figures.

1. In a 4-year study of a transitioning organic and a conventional farm in southwestern Iowa from 1998 to 2001, the conventional farm used a corn-soybeans rotation while the organic farm used various rotations of corn, soybeans, oats and alfalfa. Composted hog manure was the organic corn fertilizer, while the conventional farm used either anhydrous ammonia or 28% urea ammonium nitrate for corn. The conventional farm used an amazing array of insecticides and herbicides, while the organic farm controlled weeds by rotary-hoeing, cultivation, flame cultivation and hand-pulling of large weeds.

During the first three years (the transition period for the organic farm), corn yields on the two farms were statistically equivalent, and weed pressure the same in the first year. Organic corn yields dropped in the second year, compared to the first, and rose above the conventional farm in the third and fourth years. Organic soybean yields were the same in the first three years and 14% higher in the fourth year. Weeds tended to be a bit worse in organic fields the last three years, but still, crop yields were quite respectable. Finally, profitability on the organic farm was much higher than the conventional farm, partly because of the high market price for organic soybeans. This research was published in *Agronomy Journal* (2004) and *American Journal of Alternative Agriculture* (2003).

2. In a 21-year study of rather small plots in Switzerland, European scientists compared strictly organic and biodynamic (no synthetic fertilizers) with conventional and partly-conventional (synthetic and manure fertilizers) systems. Over two decades (1978–98) the average of all crops yielded 20% lower in the two organic systems, but winter wheat was nearly as high (90%) as conventional. But this yield performance was considered very good since fertilizer and energy used were 34 to 53% lower on organic plots, and pesticide use 97% less. Thus the organic systems were considered more productive and energy-efficient per unit of crop. Beneficial soil organisms (earthworms, fungi, bacteria) were much more abundant in organic plots and were credited with supplying a good deal of crop nutrients. The authors concluded

that organic systems are a "realistic alternative" to conventional systems. The research was reported in *Science*, 2002, Vol. 296.

3. A 6-year study (1978–83) by the University of Nebraska compared several systems, ranging from conventional continuous corn to crop rotation with manure for fertilizer. The organic system won hands down, earning a net income of $125.95 per acre, compared to $94.58 for continuous corn. The per-acre input costs for 1983 were $106.20 for continuous corn, while the rotation cost only $59.92 (or $71.52 if oat straw was sold rather than being recycled). The study was published as staff paper no. 14, University of Nebraska College of Agriculture, Department of Agricultural Economics, 1984.

4. In 1999, a year of extreme drought in Pennsylvania, the Rodale Institute compared the performance (corn and soybean yields) of conventional, legume-based organic and manure-based organic farms. The drought performance of corn under legume-based organic management was only 38% of the yield of the conventional system, but manure-based organic corn was 37% greater than conventional. Organic soybeans out-yielded conventional in both systems, 96% greater for legume-based and 52% greater for manure-based. In the four years before 1999, the two organic systems consistently out-produced the conventional system (published in *American Journal of Alternative Agriculture*, 2003).

The future? Strictly organic agriculture and complete reliance on synthetic fertilizers and pesticides are at opposite ends of the agricultural spectrum. Although initially, 50 years ago, "chemical" farming seemed to be a messiah, conditions now clearly show that it has a troubled future. The environmental damage and sickness it causes are just too great. Crop and animal quality are sinking lower. Pest and weed problems are increasing, not disappearing. Economically, it cannot at all be defended. It uses non-renewable and increasingly expensive resources, oil and natural gas, as its major input and gives us devastating ecological and health damage as by-products. The sun is setting for totally synthetic-based agriculture.

Organically oriented agriculture, or the natural-systems approach which we promote, relies on the established *natural systems* and *cycles* around which the natural world is built. Nutrients are *recycled*, not made from oil by chemical companies. The beneficial soil organisms are fostered to help make nutrients available to plants, to fight diseases and pests, and to keep soil in good condition. Only natural-systems agriculture really makes sense.

But blind reliance on total non-synthetic chemical ("organic") methods, while much better than drenching the landscape with toxic substances, can be incomplete and can be producing poor quality crops from out-of-balance soil. A fear of anything "scientific" is like sticking your head in the sand. We need to use whatever is good from modern science and technology. The tricky part is sifting out the good from the bad. This is where a sound store of knowledge about natural systems and a lot of common sense are needed. If limited use of certain synthetic fertilizer materials, along with promoting humus formation and soil life, produces superior results and does not harm natural systems, then why not follow that method?

The totally organic farmer is certainly not old-fashioned and inefficient by not using the latest pesticides, nor is he a wild-eyed radical. He is on the right track and is far ahead of most of his "progressive" neighbors.

Chapter 16

Shoo! —
PEST & WEED CONTROL

"DANGER! May be fatal if swallowed, inhaled or absorbed through the skin. Rapidly absorbed through the skin. Repeated inhalation or skin contact may, without symptoms, progressively increase susceptibility to poisoning. Do not get in eyes, on skin, on clothing. Do not breathe dust. Do not contaminate food or feed products. Keep out of reach of children. Keep out of reach of domestic animals. Not for use or storage in or around the home. This product is toxic to fish, birds and other wildlife. Keep out of any body of water. Do not apply where runoff is likely to occur. Do not contaminate water by cleaning of equipment or disposal of wastes."

T hese frightening words are quoted from the label of a commonly used soil insecticide manufactured by one of the major chemical companies. Does it sound like something you want to be around—or your family? And to think that thousands of tons of this and equally toxic chemicals are sprayed, dusted, and spread every year all over this country and most other agricultural lands, sends chills up my spine. How about yours?

Questions. Are our crops and animals *always* attacked by diseases and pests? Do weeds *always* infest our fields? Are toxic pesticides the only way of fighting these enemies? What *really* causes these problems, anyway?

Answers. We are more familiar with how our bodies work. We know that we have a built-in *immune system* which protects us from diseases and germs—well, most of the time. Actually, we have several different protective mechanisms, including our skin and mucus membranes as protective barriers, and internal antibodies and white blood cells as the defensive army. You may have heard of the substance interferon which can defend you against virus attack.

Stress. But you also realize that your natural defenses can break down. If you don't get enough sleep or exercise, or don't eat a nutritious, balanced diet, you are likely to come down with a cold or the flu. These *additional stresses* overtax the body's functions and ability to resist disease germs, and the germs, which were there all the time, are *allowed* to get a foothold. They only attack and cause the disease symptoms when the body is stressed and not in top-notch health.

The key. That is the key to understanding and preventing pest and disease attack. Not all animals and not all plants are attacked by insects or diseases. Sometimes one field will be devastated while the field across the fence remains perfectly healthy. Sometimes several plants in a row are healthy while the rest are attacked.

Both plants and animals have natural mechanisms of resistance to diseases and pests. If they have no serious genetic defects, *it is only when crops and animals are under severe stress that the diseases and pests are able to successfully attack.* They weren't in good health anyway, even before the germs or pests invaded. They were stressed to such a degree that their internal functions were beginning to be abnormal or inadequate. They were already sick (subclinically sick, to use the technical term). They would not have produced good quality crops or meat or milk. They were inferior.

Why diseases and pests? If you look at the over-all design of nature, the diseases and "pests" really have a useful function. *They are here to eliminate the sick, the weak, the unfit.* This is more easily understood by looking at a "wild" ecosystem, where many species of plants and animals live together and keep each other in check. Predators such as wolves can only catch and eat those animals that they are able to catch. Generally it is the sick or lame deer or moose that becomes a meal.

The reason pest and disease problems are so prevalent in agriculture today is that man inadvertently does things that put additional stresses on his crops and animals. He uses fertilizers and chemicals that upset soil ecology and prevent plants from getting the nutrients they need. He gives animals nutrient-deficient feed or confines them in poorly-ventilated buildings. He plants hybrid varieties that do not build high quality protein. Sometimes maybe it isn't inadvertent; sometimes it is done deliberately in the name of efficiency and profits. In any case it is short-sighted and unwise.

Weedy corn.

Pests and diseases actually provide a valuable service to us. Besides eliminating the unfit, they show us something is wrong, if we have the wisdom to heed their warning.

Many experiments have shown that insect pests and diseases attack crops that were given out-of-balance fertilization, such as too much nitrogen, or not enough phosphorus, potassium, or minor elements.

Weeds too. The same basic principle applies to weed problems. In general, most weeds grow best on poor, out-of-balance soil. That is their function in nature, to cover bare spots and waste places and prevent erosion. When they die they help build humus. Their roots bring up minerals (especially trace elements) from the subsoil and can break up a hardpan and open up tight soil. Some of them even make good forage for livestock. Weeds are hardy "opportunists," always ready to colonize poor soil.

Some weeds grow best on clay (quackgrass), others on sand (sandbur), some on acid soil (sorrel, dock), and others on saline soil (foxtail barley). Some take over on tight, anaerobic, or waterlogged soil (foxtail, bindweed, velvetleaf [buttonweed], Johnsongrass, jimsonweed). Thus weeds can be indicators of problem soils. Sheep sorrel is a classic indicator of low-lime soil. Dandelion and plantain grow well in alkaline soil, while common sunflower and giant ragweed (horseweed) do not. Many weed seeds are "programmed" to germinate in conditions of wet or anaerobic soil, and toxic decomposition of organic matter. A few weeds such as Lamb's quarters and redroot (rough) pigweed grow best on good soil, and purslane prefers fairly good soil.

Prevention or cure? When you get sick, you take whatever medicine is necessary to make you feel better. Actually the medicine does not cure you—your natural body processes do—but the medicine, plus rest and proper diet, assist the body to cure itself.

When your crops or animals get sick or are being ravaged by pests, you may have to resort to emergency measures and spray pesticides or have the vet give a drug. Rescue chemistry, it's sometimes called. You have to get a crop that year or go bankrupt. It shouldn't be that way, but those are the realities of modern agriculture.

But is that the best approach? Isn't it much better to *prevent* the disease or problem from occurring in the first place? Of course it is. It saves yields, trouble, and money. And that's just the beginning. Now some people's idea of prevention is to spray the poisons ahead of time, *just in case* the bugs come, or because the weed seeds are going to sprout anyway. But this is

Indicator Weeds

Here is what some of the common U.S. weeds tell about soil conditions (scientific names are in parentheses):

- **bindweeds** *(Convolvulus)*—crusted, tight soil, low humus
- **broomsedge** *(Andropogon virginicus)*—depleted, oxidized soil, low in calcium and possibly magnesium; poor soil structure; possible overuse of salt fertilizers
- **foxtail barley** *(Hordeum jubatum)*—wet soil, possibly high salts and low calcium, compaction, possibly acid, unavailable potassium and trace elements
- **common burdock** *(Arctium minus)*—high iron, acid, low calcium; also grows on high gypsum soil or from excess use of dolomite lime or ammonium sulfate plus lime
- **cheat, chess** *(Bromus secalinus)*—wet, compacted, puddled (fine particles, no crumb structure)
- **chickweed** *(Stellaria media)*—high organic matter at surface, low mineral content
- **chicory** *(Cichorium intybus)*—fairly good soil, clay or heavy soil
- **cocklebur** *(Xanthium pennsylvanicum)*—fairly good soil with high available phosphorus, but may have low available zinc
- **crabgrass** *(Digitaria sanguinalis)*—tight, crusted soil, low calcium, inadequate decay of organic matter
- **dandelion** *(Taraxacum officinale)*—low calcium, organic matter not decomposing
- **dock** *(Rumex)*—wet, acid soils
- **fall panicum** *(Panicum dichotomiflorum)*—anaerobic, compacted soil
- **foxtail, giant foxtail** *(Setaria)*—tight, wet soil, possibly high magnesium; seed germinates in anaerobic conditions (high carbon dioxide)
- **horsenettle** *(Solanum dulcamara)*—crusted soil, low humus
- **jimsonweed** *(Datura stramonium)*—improper decomposition of organic matter (fermentation)

- **Johnsongrass** *(Sorghum halepense)*—depleted soil, low organic matter, low calcium, possibly high iron
- **Lamb's quarters** *(Chenopodium album)*—rich, fertile soil; good decay of organic matter, high humus
- **common milkweed** *(Asclepias syriaca)*—good soil, generally grows in fallow areas
- **mustards** (wild mustard, yellow rocket, wild radish, peppergrass, etc.) *(Brassica, Raphanus, Lepidium)*—crust, hardpan, poor soil structure, poor drainage; high potassium fertilization
- **nettles, stinging nettle** *(Urtica)*—anaerobic, toxic soil, wrong decomposition of organic matter (fermentation)
- **pigweeds** *(Amaranthus)*, redroot (rough) pigweed *(Amaranthus retroflexus)*—good soil
- **purslane** *(Portulaca oleracea)*—fairly good soil
- **quackgrass** *(Agropyron repens)*—wet, anaerobic soil, high aluminum (toxic); in West, low calcium and high magnesium and sodium
- **red sorrel, sheep sorrel** *(Rumex acetosella)*—acid soil, low calcium, low decomposition of organic matter
- **Russian thistle** *(Salsola kali* var. *tenuifolia)*—salty soil (high sodium and potassium), low calcium and iron, low organic matter
- **smartweeds** *(Polygonum)*—wet, poorly drained soil
- **thistles** *(Cirsium)* & **sowthistle** *(Sonchus oleraceus)*—fairly good soil
- **tumbleweed** *(Amaranthus albus)* (Russian thistle is also called a tumbleweed; see above)—dry soil, low humus
- **velvetleaf** (buttonweed) *(Albutilon theophrasti)*—anaerobic soil, wrong decay of organic matter (fermentation)

Compiled from: C. Walters, 2003, Eco-Farm—An Acres U.S.A. Primer; *E.E. Pfeiffer, 1981,* Weeds and What They Tell; *S.B. Hill & J. Ramsay, 1977,* Weeds as Indicators of Soil Conditions.

only a band-aid approach, attacking the *symptoms* of a problem and not the *real causes*.

Diseases, weeds, and pests can really be prevented and the problems solved by eliminating what causes them. Diseases and pests attack when plants or animals are under too much stress and their natural defenses are down. Weeds are a problem on anaerobic and out-of-balance soil. Before we go into detail on solving these problems, let's look at the results of the other approach—the use of herbicides and pesticides.

What chemicals do. What do toxic chemicals actually do? How do medicines work? Do they have any other effects besides killing germs, bugs, or weeds?

The basic purpose for using pesticides and medicines is to kill, whether it is germs, weeds, or bugs. Therefore these chemicals are usually at least toxic or poisonous to the "target organisms"—those species we want to kill.* A few are pretty specific and only kill one or a few species, but the great majority can kill other non-target organisms. So in the case of insects, insecticides not only eliminate, say the greenbug aphid, but also the predators and parasites that help keep it under control—if they are given a chance. Green lacewings, ladybugs, and tiny parasitic wasps are efficient natural enemies of aphids. Or how often does herbicide drift or carryover from last year injure a crop? These are undesirable side effects.

Many medicines (drugs) are pretty specific for a certain type of pathogen, but like pesticides, they have side effects. They disturb or inhibit some body function. They put additional stress on the body, which can weaken an animal so much that it comes down with another disease.

Insecticides kill insects in a variety of ways. Some have to be eaten to kill (stomach poisons), some act when breathed into the respiratory system (fumigants), and some kill when any part of the insect's body touches them (contact poisons). The actual death of the pest can be caused by cell destruction of certain internal tissues. Others attack the nervous system. DDT causes nerve cells to "fire" repeatedly. Organophosphate insecticides such as parathion were developed from nerve gas research

---

Actually, there are a few pesticides that kill "bugs" but are not poisonous, such as diatomaceous earth, which abrades insects' outer cuticles and they die from drying out; or attractant odors that lure insects into traps; large raindrops will kill aphids just by mangling them.

There are several terms used to name toxic chemicals used for pest control. The most general would be *pesticide*. Specific types of pesticides would include *herbicides* to control weeds, *fungicides* for fungi, *insecticides* for insects, *acaricides* or *miticides* for mites, *nematicides* for nematodes, and *rodenticides* for rodents and other small mammals.

during World War II. They suppress vital enzymes, cholinesterases, which regulate nerve impulses.

Herbicides kill plants by such mechanisms as interfering with various cellular metabolic functions (amino acid production, energy transfer, cell membrane functions), destroying photosynthesis capability, decreasing stability of the cell's nucleus, and interfering with cell division. The well-known 2,4-D acts as a synthetic plant hormone, resulting in abnormally slow tip growth and sometimes tumors.

Side effects. Many herbicides and fungicides are absorbed into the plant's tissues and carried (translocated) throughout the plant. They can end up deep in the soil by going through the roots. They can enter crop plants and end up in animal feed and the fruits and vegetables we eat. In vineyards and orchards, where fungicides containing heavy metals are used frequently, copper, zinc, and manganese levels have been increasing in apples and grapes. Apples from sprayed orchards in Europe contained 3.5 times the zinc, 2.4 times the copper, and 5 times the manganese as unsprayed apples. Most pesticides are toxic to at least a few plant species, and soil fumigants are toxic to most plants. If crops are planted too soon after application, if soil is cold and wet, or if pesticide residues have built up over the years, crop failure can result.

Pesticides can affect the crop plant's functions, leading to abnormal levels of nutrients, plus lower quality and flavor. In a study on beans, 2,4-D caused lowered protein content, while in sugar beets it caused such a high nitrate content that they would be toxic if eaten by cattle. The soil insecticide benzene hexachloride (BHC) causes off-flavors in many fruits and vegetables, as does lindane. Some fungicides and insecticides used on cherries cause poor flavor and quality (color, size, weight, acidity).

Almost nothing is known about the effects of combinations of two or more pesticides. In some cases a combination is much more toxic than either chemical is alone (called a synergistic effect). The nerve poison aldicarb (TEMIK) is an extremely toxic insecticide (a few drops on your skin can be fatal). In the environment, it rapidly oxidizes to form aldicarb sulfoxide, which is 47 times more toxic! Dr. Donald Kaufman, a USDA microbiologist, found that in the soil, certain pesticides block the breakdown of other pesticides. Aldicarb is among the chemicals that have been found to act that way.

Many herbicides and insecticides are volatile, evaporating quickly into the air. Others evaporate more slowly, but on a hot day the rate speeds up greatly. This may be good in that it reduces soil contamination by poisons, but does it create a toxic air pollution and a deadly "poison rain," similar to acid rain, miles away? Evidence indicates that it does. Pesticide evaporation, plus airborne spray particles and dust can literally be carried around the world by upper air currents. Atrazine, 2,4-D, DDT, DDE, BHC, and dieldrin have been found in rain in such places as England and Hawaii. Extremely high levels of aldrin, chlordane, toxaphene, and DDT were measured in the air over agricultural areas of Florida and Mississippi in the 1960s. When the volatile thiocarbamate herbicide EPTC is applied to soil in water solution, 70% of it evaporates the first day if the soil is moist (45% if it is dry).

Soil problems. When pesticides get into the soil (some are applied directly to or into the soil, some drip off plants or are washed into the soil by rain), many undesirable and unexpected results can occur. Some soil organisms are killed immediately, while others are fairly tolerant.

The insecticide heptachlor is toxic to nitrifying bacteria at 50 ppm and legume nitrogen-fixing bacteria at 100 ppm. Carbofuran at 4 lbs./acre killed 95% of the earthworms in one study, and disulfoton used for aphids on cotton at 2 lbs./acre in the furrows killed 95% of certain soil arthropods (collembola and mites). Diazinon at 8 lbs./acre killed 64% of beneficial predatory ground beetles, while parathion at 8 lbs./acre killed 95% of them.

Fungicides and soil fumigants are the worst, killing large numbers of soil organisms. It may take anywhere from weeks to up to a year before sterilized soil is recolonized by normal populations of organisms. Plant diseases can be worsened if a fungicide kills off beneficial microbes that normally hold the pathogenic species in check, although in some cases, beneficial disease-fighting species proliferate after soil treatment. Pesticides can cause mutations of some soil organisms (*Aspergillus* fungi).

Most pesticide studies are only concerned with short-term effects. In one long-term study, yearly applications of 3.56 lbs./acre of atrazine to kill weeds in an orchard caused significant soil changes after 14 years. Humus content of the weed-free soil had dropped to nearly half that of untreated soil, and pH was lower (4.7 in atrazine soil, 5.3 in untreated). Total numbers of soil bacteria and fungi were about the same, BUT the balance was greatly upset: some humus-forming bacteria and anaerobic species were permanently reduced, while other humus-formers and denitrifying bacteria were only temporarily reduced by atrazine. Other types of microbes (ammonifying and nitrogen-fixing) were increased. The normal nitrogen cycle in the soil (nitrification: change of ammonium to nitrate nitrogen) was seriously reduced.

Herbicides, fungicides, and soil fumigants have been found to release plant nutrients in the soil by killing some microbes and stimulating others, from breakdown of the pesticide, and by direct chemical action of the pesticide on the soil. If the plants can't use all of the released nutrients, some can be leached away. Some trace elements (manganese, copper, zinc) can be made more soluble to such an extent that they become toxic. High levels of ammonia can build up to toxic levels (to sensitive plants) because of the high numbers of dead soil microbes and inhibition of nitrifying bacteria.

Although most pesticides apparently do not harm soil tilth directly, the killing of humus-producing organisms and the reduced amount of plant residues returned to the soil (from herbicides) can lessen humus levels and degrade soil structure in the long run. Also, the high levels of ammonia sometimes produced can destroy tilth more quickly.

Following application of soil fumigants, fungicides, or other treatments that kill soil organisms, crop growth is often inhibited, and plants do not take up enough phosphorus, zinc, or copper. The problem may be caused by toxins produced by excess growth of certain microorganisms, while beneficial species that would have destroyed the toxins have been eliminated. Similarly, disease pathogens can get the upper hand because their natural enemies have been destroyed. Beneficial mycorrhizae fungi are killed by many fungicides.

Although not sold as a pesticide, a chemical (nitrapyrin) called a nitrogen stabilizer is applied to soil to inhibit the nitrification process by killing *Nitrosomonas* bacteria. Thus ammonium is not converted to nitrite (and then to nitrate), and nitrogen stays in the soil for a longer time. Research has found that nitrapyrin and its breakdown product, 6-CPA, have relatively low toxicity for most other forms of life in soil, water, and on land (although few species have been tested). At high concentrations (100–1000 ppm) soil fungi are inhibited, however. If soil and weather conditions are not right, there may be no crop yield increases, according to an article in *Down to Earth* (1976, Vol. 32, No. 1, p. 22–26).

In the ecosystem. The problems that result when pesticides become distributed in the environment are well known. Very small amounts end up in streams and lakes, get into algae, and then into crustaceans that eat the algae. Next into fish and then into birds and mammals. Then into humans, which adds to the pesticide residues we get directly from the vegetables, fruits, and meat we eat. At each step up the food chain, the amount of pesticide is concentrated, so while lake water may only contain 0.000001 part per million DDT, crustaceans may have 0.001 ppm, fish and birds 2.0 ppm, and humans 6.0 ppm (see diagram, p. 238). Nearly every person tested by the Department of Epidemiology, Miami University, Miami, Florida, had DDT, dieldrin, lindane, and other insecticides in their body fat. Also, 100% of the people had pentachlorophenol in their urine, a by-product of pesticides which causes fetal tumors in rodents.

Dr. Americo Mosca of Turino, Italy, has found that toxic agricultural chemicals cause damage to tissues by ionization, leading to mutations, birth defects, and cancer, just as radiation from atomic fallout can do.

Resistance. Probably the most ironic side effect of the use of pesticides is that the very pests they were supposed to eradicate have bounced back even stronger than before. They have developed resistant varieties, which are immune to the old pesticides, and the chemical companies are kept scrambling to come out with new, stronger poisons. We now have "superflies" and potato beetles that have to be sprayed weekly rather than once or twice a year. The number of insects resistant to one or more insecticides has shot up from about 25 in the mid-1950s to over 500 in 2000. There are more than 130 weed species resistant to herbicides. Since 1970, over 100 species of fungicide-resistant plant diseases have been found. The problem is occurring world-wide.

Herbicides can even worsen insect problems. In one test of the effect of 2,4-D use on corn, 16% of untreated plants had corn borers, compared to 28% and 24% for normal and 1/4 normal doses of 2,4-D. Numbers of corn leaf aphids per plant were 618 (untreated), 1679 (normal dose), and 1388 (1/4 dose).

Some fungus diseases are increased by application of herbicides; early blight of tomato and mosaic virus of tobacco, cucumber, and cotton are increased by 2,4-D, while damping-off of cotton increases when trifluralin is used.

Conclusions. One over-all observation that can be made about the use of synthetic chemicals as a "quick fix" for agricultural problems is that *they interfere with nature.* They do not work with natural laws and processes—they try to overpower nature. And as we are finding out, it just doesn't work!

We are going to have to swallow our pride and admit we aren't as smart as we thought we were. We need to study nature more closely, discover (or rediscover) its laws and processes, and put them to work for us. That's how things were intended to work.

Pests. Again, why are some animals and plants pests, anyway? Pests are defined as animals or plants that are harmful to ourselves or our property; so as applied to agriculture, this would include the insect, mite, and nematode pests of crops and livestock, as well as weeds, which are considered to be the worst pests of crops, *and* those diseases of crops and livestock which are caused by microorganisms (viruses, bacteria, and fungi). Nutrient deficiency diseases would not be included, and many experts exclude animal diseases and parasites, although we will include these animal pests.

These pests can cause harm by either feeding directly on crops, stored crops, or animals; by releasing toxic substances; by competing with plants for water, nutrients, or sunshine; by carrying plant and animal diseases; and by weakening a plant or animal and allowing other pests or diseases to get started. The number of pest species is relatively small (less than 1% of the total number of insect species are pests), but the few pest species can sometimes be very common and destructive. In fact, many insects, mites, nematodes, fungi, and bacteria are very beneficial because they eat, parasitize, or cause diseases of the pest species. Even some "weeds" are used for forage in some parts of the world, and if no crops are growing on the land, weeds help prevent erosion and nutrient leaching. In addition, all organisms in nature are part of the food web, or "balance of nature," in which one species uses another for food; so indirectly, everything has some value.

If man had not upset the balance of the food web in so many parts of the world, all or nearly all of the so-called pests would not be pests, but would be held in check by their natural enemies. Most of the worst pests have been carried by man from their original homeland, where their natural enemies controlled them, to some other part of the world, where few if any natural enemies exist. Often their names show this, such as the European corn borer, Japanese beetle, and the Mediterranean fruit fly in the U.S., and the Colorado potato beetle, which has wreaked havoc in Europe.

When is a pest a pest? Is all damage that pests do really important? Should we spray a whole field with poison just because of a few leaves being nibbled? Actually, one-half or more of insecticide spraying is unnecessary. Speaking in terms of immediate dollar loss, if the damage a pest does to a crop or livestock does not exceed a certain level, called the economic threshold, control methods do not pay. Not all pest damage causes serious harm. Most plants can lose up to 30–40% of their leaf area before they are harmed, and hardy plants can outgrow some types of damage. Sometimes leaf damage does not occur until after the grain or fruit is mature, so it doesn't matter. Healthy livestock and poultry can tolerate a certain level of parasite attack. More important, healthy plants and animals can resist and ward off diseases and pests. It has been found as far back as 1961 that the basic cause of corn stalk rot is not the attacking fungi, but low sugar production caused by plant stress, including drought and cloudy weather. These stresses "curtail a plant's ability to produce enough sugar to supply the food needed to sustain plant vigor . . . the stalk and roots get shortchanged . . . fungi that normally are warded off invade and destroy the tissues . . . weakening [stalks] and leaving them susceptible to lodging." Balanced soil fertility is one of the recommended ways to reduce stresses (*Successful Farming*, Nov. 1980, p. 36–AD). We will have more to say along these lines later.

Pest control philosophies. Unfortunately, many people's pest control philosophy can be summed up by, "The only good bug is a dead bug!" If you don't believe this, just think of the wild goose chase the pesticide companies have led us on for the last 50-plus years. More and stronger insecticides and herbicides. But pest problems have only gotten worse, for the most part, and the ecological side effects have been devastating. *Finally*, now that it is almost too late, some of the "experts" are starting to preach another philosophy.

Pest *control* (also called pest *management*) does not mean pest *elimination*. There is no way every single corn earworm can be killed (more would come in from the neighbor's fields anyway!) or all damage can be eliminated. Thus, in pest control, the main goal is keeping damage below the economic threshold, if possible. There are two ways of doing this: (1) prevention of pest attack in the first place, and (2) control of the numbers of pests, to reduce the intensity of attack. Obviously the first is the most desirable, but both are usually necessary.

There are many ways of attacking pests, some widely used, some largely impractical at present, and some still in the research stage. Using poisonous pesticides is only one, and usually the least desirable in the long run. In modern pest control, often called *integrated pest control*, we attempt to use more than one control method, to attack the pest's weakest points, and to do so in an ecologically sound way. This requires detailed knowledge of the pest's life cycle and weak points, knowledge of when and how to use different control methods, plus frequent surveys of fields to obtain up-to-date information on local pest abundance.

Control methods. We can divide pest control methods into two large categories, (1) natural control, referring to weather factors and natural enemies of pests (predators, parasites, and diseases); and (2) artificial (applied) control, which refers to any human-directed methods. There are many artificial control methods; some of the more commonly used ones include (a) using human-released natural enemies (such as lady-bird beetles or a bacterial caterpillar disease or parasitic wasps), (b) releasing sterilized males of certain insect pests (such as the screw-worm fly and mosquitoes), (c) using attractant or repellent chemicals or plants (such as sex-attractant odors in an insect trap or spraying garlic oil on your garden plants), (d) planting resistant or repellent varieties of crop plants, (e) cultivation methods that kill weeds and hibernating insect pests, (f) removal of dead plants and spoiled fruit to prevent disease spread and eliminate insect hibernation sites, (g) adjusting planting and harvesting times to escape major pest attack, (h) crop rotation to escape soil-harbored pests and diseases, and (i) using toxic chemicals (poisons) to kill pests.

There is a wide variety of toxic chemicals that are used for pest control, ranging from some that do not cause serious ecological problems to others which are incredibly dangerous poisons related to wartime nerve gases. The least dangerous general group of poisons is often called the *botanicals*; that is, natural chemicals produced by plants. They break down soon (they are biodegradable) and seldom pollute the environment. Commonly used botanicals include pyrethrum (also called pyrethrin), rotenone (this one is toxic to fish), and sabadilla. A second group of toxic pesticides is the *inorganics*, some of which have been used for hundreds of years. There is a wide variety of chemicals in this group; some are relatively harmless, such as sulfur and lime-sulfur, but others are very toxic, such as copper sulfate, arsenic compounds, and mercury compounds. The third group

When corn prop roots are reluctant to enter the soil, herbicide carryover is to be suspected.

of toxic chemicals is the recently developed (since World War II) *synthetic organics*, most of which are very toxic and may not break down rapidly in the environment (they are called persistent pesticides). Fortunately, many can eventually be broken down by sunlight, oxygen, and bacteria, but some contaminate the entire ecosystem for years. Some have been found all around the world, including in the middle of the oceans, and in the bodies of Eskimos and polar bears. Even human embryos and nursing mother's milk carry pesticide residues. When fish or birds of prey eat food contaminated with pesticides, the poisons can become tens or thousands of times more concentrated in their bodies. Typical amounts of DDT measured in parts per million found in the environment are (from C.A. Edwards, *Persistent Pesticides in the Environment*, 1970, p. v):

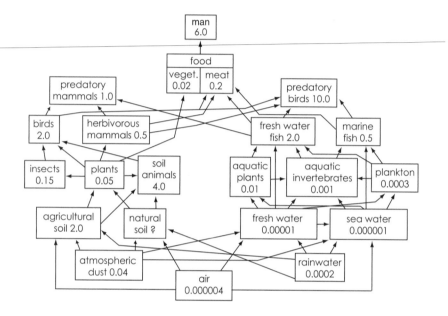

Solutions or problems? The use of toxic herbicides and insecticides has been, and still is (just read the ads in any farm magazine), proclaimed as the savior of agriculture. USDA and state university pamphlets, and even some Secretaries of Agriculture have painted a picture of starvation and doom if chemical pest control is not followed, using such figures as . . . without chemicals crop production would fall 30% and food prices would rise 50–70%. But what have been the results of over 30 years of heavy use of toxic pesticides? Consider these facts:

1. DDT, DDE, and dieldrin were found in the fat tissue of 100% of 54 cadavers tested by researchers in Dade Co., Florida. Those who were heavy users of household pesticides had over four times the DDT and nearly three times the DDE than those who were low users. The levels of DDE and dieldrin were much higher in those people who had died of brain disease and cirrhosis of the liver (*Chemical Fallout*, 1969, p. 301, 308–309).

2. After the herbicide 2,4,5-T (Agent Orange) was used for defoliation during the Vietnam war, birth defects increased greatly in the native population (H. Wellford, *Sowing the Wind*, 1972, p. 195–196). In 1970, drift from U.S. Forest Service spray in Arizona caused cases of stillborn children and deformed goats and chickens. Tests run for the National Cancer Institute found that rats produced 39% abnormal fetuses at a low dose and 90–100% at "relatively high doses" (R. Harmer, *Unfit for Human Consumption*, 1971, p. 93). A contaminant of 2,4,5-T is the chemical dioxin. In 1976, an explosion at the Icmesa chemical plant sent a cloud of dioxin over the northern Italian town of Seveso. About 700 people near the plant were evacuated. Within weeks, hundreds of villagers developed a skin disorder, chloracne. In 1979, the number of women who had been exposed to dioxin that died of breast cancer was double the national average. The number of babies born with deformities rose from 3 in 1975 to 53 in 1978. Farming was banned on 4500 acres. In 1982, the Italian government lifted the ban on farming in some areas (soil tests showed no "toxicity"), but in order to reopen the road from Seveso to Meda, the soil along the road had to be turned over. The government won't release details on soil and medical tests (*Newsweek*, May 10, 1982, p. 14, 18).

An herbicide spill.

3. In 1961, an organic phosphate insecticide, phosdrin, accidentally spilled on a bundle of blue jeans being trucked from Fresno to Los Angeles, California, in the same truck. Six children who wore the jeans were stricken with such symptoms as rapid heartbeat, irregular breathing, glassy eyes, muscle twitching, diarrhea, nausea, vomiting, and abdominal pain (R. Harmer, *Unfit for Human Consumption*, 1971, p. 21–23). Few farmers or homeowners read the labels before using pesticides. Safety standards are commonly violated. Farm workers and fruit pickers are often sprayed by planes. Sixteen out of 24 vineyard workers were overcome when they entered a vineyard in California 33 days after it had been sprayed with an organic phosphate (nerve gas type) insecticide (the permissible time limit in California was 14–28 days). Vegetable farmers anxious to deliver flawless produce to market have been known to douse their crops with incredible amounts of insecticide; one lettuce grower was caught using chlordane, endrin, dieldrin, DDT, toxaphene, malathion, cryolite, and rotenone only 11 days before harvest (B. Mooney, *The Hidden Assassins*, 1966, p. 106). The U.S. Food and Drug Administration's budget and staff are so small that only a tiny percentage of interstate food shipments can be inspected. Food from foreign countries can have high levels of pesticides.

4. U.S. export trade is sometimes harmed because many foreign countries no longer accept pesticide-contaminated U.S. food and tobacco. The European Common Market countries set much lower pesticide residue tolerance limits than the U.S.

5. *Combinations* of toxic chemicals are often more dangerous than single ones (this phenomenon is called synergism). Combinations of chlorinated hydrocarbon insecticides at "harmless" levels caused liver damage in test animals, but did not when administered singly at the same dose (*Quarterly Bulletin of the Association of Food and Drug Officials of the U.S.*, Fall 1950, p. 90). The insecticide malathion, considered a "relatively harmless" organic phosphate, became over eight times more toxic when administered to animals along with EPN (*Journal of Pharmacology*, Vol. 121, p. 96).

6. About 500 species of pests, insects and mites have developed a resistance to pesticides. "Superflies" can tolerate a dose equivalent to feeding 14 pounds of poison to a 200 pound man (R. Harmer, *Unfit for Human Consumption*, 1971, p. 58). The German cockroach is so resistant to both chlorinated hydrocarbons and organic phosphates that it has been called "indestructible" (*Wall St. Journal*, Feb. 14, 1968).

7. Pesticides kill many organisms in the ecosystem, not just the "target" pests. Aldrin and dieldrin greatly decrease populations of soil arthropods, while endrin kills many earthworms (*Residue Reviews*, 1975, p. 40–41). Pesticides can affect the metabolism of plants, and plants can accumulate pesticides from the soil (*Residue Reviews*, 1974, p. 73–74). Over 30 currently used pesticide preparations are labeled as toxic or highly toxic to honeybees (Iowa State University Entomology Extension sheet). Some pesticides are very stable and persist in the environment for many years (others break down rapidly); for example, after 14 years the following percents of pesticides were still present in a sandy soil: aldrin 40%, chlordane 40%, endrin 41%, heptachlor 16%, dilan 23%, isodrin 15%, BHC 10%, toxaphene 45% and (after 17 years) DDT 39% (*Residue Reviews*, 1974, p. 109–110). Persistent pesticides are passed along in the food chain and are *concentrated* as they go: in Clear Lake, California, small amounts of DDD (14 parts per billion) were used for gnat control. The microscopic plankton organisms in the water were found to have 5.3 parts per million (ppm) DDD, plant-eating fish had 40–300 ppm, carnivorous fish had 2,500 ppm, and birds had 1,600–2,000 ppm (R. Harmer, *Unfit for Human Consumption*, 1971, p. 62).

This list of "horror stories" could go on and on, but I think you get the point: the use of pesticides is not ecologically sound. William Longgood sums it up well in his book, *The Poisons in Your Food* (1960, p. 255): "The fact that sprays are claimed to be necessary to grow food is not a warrant for their increased use; it is an indictment of the critical plight of American agriculture and the waning fertility of our once-rich soils. Poison sprays, rather than being a solution, are merely a delaying action, a final desperate maneuver to stave off catastrophe."

There HAS to be a better way.

Nature's way—"bugs." The ultimate cause of insect, mite and nematode pest problems, and plant diseases is out-of-balance soil, for in general these pests attack unhealthy, sick plants and leave the vigorous, healthy ones alone; and of course, the main cause of sick plants is "sick" soil. It may seem hard to believe that insects and diseases attack unhealthy plants and leave the healthy ones alone, for some farmers' crops and gardens never seem to escape attack, but there are many farmers and gardeners who follow ecologically sound methods and can vouch that it is true; and scientifically performed experiments also prove it. Having had a traditional university education in entomology, I also could not believe it until I saw it.

Accounts of armyworms detouring around one farmer's field of oats and destroying his neighbor's, or of one field of corn being attacked by corn borers but not another, are not just "fish stories"—they are true. I have seen one field of oats in healthy condition, while all those around it were being decimated by aphids and red leaf disease in 1982. What could make the difference between two neighboring fields, one decimated by insects or disease and the other untouched?

Natural plant protection. Let's take a look at how nature protects plants from diseases and pests. First, you should realize that disease pathogens and pests, in general, are always around, at least in the warm months. Yet not all plants come down with diseases or are attacked by pests. So diseases and pests are not inevitable, and the germs and bugs are not the real causes of the problems.

As we have said earlier, they are just there, ready and waiting, to eliminate any inferior, unfit, unhealthy plant which is weakening under stresses. Food from that plant would not be good for man nor beast, so the pests and pathogens are trying to do us a service by eating them before we can. Most of us, however, do not understand this and zap them with poison and eat the inferior food anyway. And then we wonder why we get sick or die of heart attack at age 45.

Pathogens and pests do not successfully attack healthy plants because, just as animals and humans do, they have various methods of repelling, resisting, and killing the attackers. New ones are being discovered every year.

Natural resistance. Methods and factors that contribute to natural resistance to diseases and pests are as follows:

1. *Barriers.* Plants are covered on the outside by a layer of waxy cuticle. It not only protects from loss of water from the cells, but if it is thick, it can protect from some pathogens. Some plants have a dense mat of hairs or spines on their leaves or stems, which can be an effective barrier to insects. In some species or varieties, the hairs have tiny hooks at the tip, or release a sticky secretion. These are perfect traps for small insects, such as aphids, leafhoppers, flea beetles, and young corn borer larvae. Research has been done to develop special varieties of hairy potatoes, alfalfa, and probably other crops.

2. *Repellent chemicals.* Some plants produce strong-smelling odors or toxic chemicals which keep pests from even starting to attack. Generally, if they smell strong or unpleasant to us, they will repel insects. Examples include turpentine and resins found in conifer trees, mint oils, citrus oils,

and garlic and onion oils. The use of marigolds, garlic, chrysanthemums, anise, nasturtiums, and other plants to protect vegetables from pests is well-known to many gardeners.

A certain species of wild potato has been discovered to produce a chemical that is similar to the odor (called an alarm pheromone) that aphids use to alert other aphids that there is danger nearby. Thus, aphids don't even come near the plants. Researchers are studying strong-smelling oils from herbs and spices for commercial use in controlling stored product pests (grain beetles and whiteflies).

3. *Phytoalexins.* If bacteria invade a leaf and kill a few cells, or if a fungus spore germinates, some plants secrete chemicals called phytoalexins, which kill the pathogens or inhibit germination of more spores. Corn produces two phytoalexins that give resistance to the northern corn leaf blight fungus. Phytoalexins have been found in dozens of species of plants and give resistance to fungi, bacteria, viruses, and nematodes.

4. *Amino acids and proteins.* Certain proteins and amino acids that are not part of proteins are found in resistant varieties of plants. Treating susceptible apple trees with the amino acid phenylalanine markedly increased their resistance to apple scab disease in one study. Both synthetic and natural amino acids can control certain diseases. For example, cysteine, cystine, leucine, and methionine inhibit cucumber scab; and serine, threonine, isoleucine, cysteine, histidine, methionine, glutamic acid, arginine, and tryptophane were effective against wheat rust in lab tests. More effective disease control results when there is a proper balance of amino acids, vitamins, and trace elements. Excessive fertilization with inorganic nitrogen salt fertilizers and urea increases susceptibility of many plants to fungal diseases, and high potassium increases late blight in potatoes because of unbalanced amino acids in the plants.

5. *Other plant chemicals.* Plant cells produce an amazing variety of chemicals as by-products of metabolism, sometimes called secondary metabolites. The functions of many of them are unknown, but some can make the plant resistant to pests or diseases. Some of them are just plain lousy-tasting to insects. Others cause insects to stop eating or reduce their fertility. Some inhibit the insects' digestive enzymes so that they are not able to digest their food. Some inhibit the growth of fungi or bacteria.

These metabolic chemicals include tannins, protocatechuic and chlorogenic acids, coumarin derivatives, proteinase inhibitors, and polyacetylenes. Some are only produced when the plant is injured or attacked by pests or disease pathogens.

A newly developed variety of sweet potato is resistant to rootknot nematodes, fusarium wilt, and ten species of insects, including wireworms, grubs, and flea beetles. A wild tomato from Ecuador has a substance in its hairs that kills nearly 100% of whiteflies, flea beetles, tomato and tobacco hornworms, and Colorado potato beetles. Chemicals found in corn, including one called DIMBOA, inhibit or repel young European corn borers and spores of northern corn leaf blight and gibberella stalk and ear rot.

6. *Genetic varieties.* It is well known that certain varieties of plants are more resistant to diseases and pests than others and that this resistance is controlled by one or more genes. Thus, modern agriculture has spent very much research effort to discover and develop resistant genetic varieties. The exact mechanism of resistance may be one of those previously discussed, such as repellent odors, barriers, phytoalexins, or other plant chemicals.

By "building in" the resistance in a variety of hybrid seed, the natural mechanism is automatically made available to the farmer. Varieties of wheat resistant to wheat rust and Hessian flies, soybeans resistant to phytophthora root rot, alfalfa resistant to verticillium wilt and aphids, and corn resistant to European corn borer are examples.

7. *Overall health of plant.* A stressed plant's functions are out of kilter. Some of the hundreds of internal cell and tissue functions will be abnormal or inhibited. Under anaerobic soil conditions such as are caused by flooding, roots do not absorb nutrients, imbalances in hormones occur, cellular metabolism shifts to fermentation, toxic levels of alcohol can build up in tissues, etc. Disease pathogens readily attack such stressed plants.

Stressed plants may not be able to produce their normal resistance-chemicals, such as phytoalexins or amino acids. Stressed plants even look and smell different to insects, which is how insects locate sick plants to attack. Fascinating research that proves this has been done by Philip S. Callahan (see box, p. 245).

8. *Soil life.* As we have mentioned in Chapter 3, several types of soil organisms directly or indirectly protect plants from diseases and pests. A whole host of beneficial organisms act together in healthy soil to produce a proper balance of plant nutrients and growth factors, thus producing a vigorous plant that will be more resistant. More specifically, certain fungi (mycorrhizae, *Penicillium*, etc.), actinomycetes, and bacteria (especially those near roots, in the rhizosphere) directly attack disease pathogens by releasing antibiotics or by mechanisms that starve them to death. Certain species of fungi actually trap and eat nematodes (see box, p. 246).

How Bugs "Tune In" to Sick Plants

The way in which insects locate the plants they intend to eat or lay eggs on (so their offspring can eat) is extremely fascinating. Most pests prefer to eat only certain types of plants, sometimes only one species; for example, the European corn borer prefers corn but will also attack sorghum and beets, while the onion maggot only attacks onions. But how does a female corn rootworm beetle tell a corn plant from a soybean at a distance of 100 feet, and even more amazing, how does she tell a sick corn plant from a healthy one? Entomologists used to assume that insects located objects by odor, vision, taste, and feel.

But a former USDA entomologist, Dr. Philip Callahan, opened up a whole new dimension in insect activity. Dr. Callahan, who has a background in electronics and radio, found that some of the tiny hairs on the antennae of insects are sensitive to varying wavelengths of infrared. Now, everything in nature radiates ("broadcasts") invisible infrared wavelengths depending on its temperature and molecular composition. Odor molecules broadcast at different wavelengths depending on their type, concentration, temperature, and vibrations. Thus insects can easily distinguish between different kinds of plants and other objects in their environment. Dr. Callahan describes these things in his book, *Tuning into Nature* (1975) and in *International Journal of Insect Morphology and Embryology*, Vol. 4, p. 381–430. It is known that sick plants have a different temperature than healthy ones (*Successful Farming*, Dec. 1981, p. 26–P), and that a sick plant with upset metabolism will produce different odor molecules than a healthy one. Thus sick plants will "look" different at infrared wavelengths to an insect. Man already uses this to survey crops by taking infrared photographs from airplanes and satellites. We are just duplicating what the insects already do.

Predatory Fungi

A science fiction writer could hardly come up with a stranger tale than the story of soil fungi that "prey" upon soil animals, especially nematodes (roundworms, threadworms), which are often serious plant parasites (not all nematodes are harmful pests; many are predators in the soil or else live on decaying organic matter; others, however, are animal parasites).

Some of these fungi produce tiny donut-like "nooses" at intervals along their filaments. When microscopic animals, including potentially harmful nematodes, poke their bodies through a "noose," it instantly constricts, trapping them; then the fungus dines on its hapless victim. A nematode seldom escapes. Other species of fungi trap nematodes on sticky knobs and branches (C.L. Duddington, *The Friendly Fungi*, 1957; G.L. Barron, *The Nematode-destroying Fungi*, 1977). One type of these fungi grows nematode traps only when nematodes are in the vicinity; otherwise it lives without "fresh meat" (W.A. Jensen & F.B. Salisbury, *Botany: An Ecological Approach*, 1972, p. 365).

Unconstricted and constricted ring traps.   *Sticky knobs and branches*

One of the best ways to increase beneficial soil organisms is to add organic matter (green manures, animal manures, compost). In one study, a California grape grower who had nematode problems, even after using the toxic nematicide, Nemagon, applied three tons of compost per acre. The nematode problem ceased, soil structure improved, and yields started up.

Tests at the University of Illinois found that inoculation of soybean seeds with certain strains of rhizosphere bacteria protected plants from phytophthora root rot just as effectively as plants treated with a fungicide and out-yielded untreated plants by 129%. Similar inoculation of potato seed pieces increased yields up to 33%, while radishes increased in root weight 60 to 144% over untreated plants.

Plant pathologist R.J. Cook believes that one way humus and soil microorganisms help control diseases is by small amounts (a few parts per million) of ethylene gas, produced by some anaerobic soil bacteria. Ethylene, which is also produced in plants, where it functions as a plant hormone, inhibits the germination of disease fungus spores (such as *Fusarium* and *Sclerotium*). The ethylene is produced especially when organic matter is added to the soil, providing food for the bacteria, while excessive tillage and high nitrate levels decrease ethylene production.

9. *Soil nutrients.* The proper amounts and balance of nutrients must be present in the soil in order for plants to be healthy. If one element is out of balance (too high or too low), others can be affected. Deficiencies can result. High or low pH can also restrict nutrient availability and cause deficiencies. Some of the interrelations of mineral nutrients and pH are shown in the diagram in Chapter 12 (p. 183).

For example, calcium and magnesium are "antagonists," "competing" for root absorption, so to speak. If there is too much magnesium relative to calcium, the plant will be deficient in calcium, and unhealthy cell walls, poor root development, slow growth, and poor crop quality can result. Increased calcium levels in potatoes reduced *Erwinia* soft rot up to 50% in one test. High potassium can cause calcium, magnesium, boron, iron, and zinc shortages; and when potassium chloride (muriate of potash) is used, production of amino acids, proteins, chlorophyll, and vitamins C, B_1, and A (carotene) can be reduced; crop quality can go down. Excess nitrogen fertilization is notorious for increasing disease and pest vulnerability in plants. *Many* experiments have proved this. For example, high nitrogen, as well as surface mulch (as would occur in reduced tillage systems) increased corn borer infestation in one study, possibly because a corn borer inhibitor chemical produced in the rind of the corn stalk is reduced.

Excess nitrogen produces a fast-growing, but weak and watery plant. In one experiment, high nitrogen fertilization decreased the numbers of soil fungi that fight diseases, while a medium level of phosphorus increased them. In an experiment on red clover and *Fusarium* diseases, vigorously growing plants had the least amount of disease. A moderate amount of nitrogen, as well as the proper amounts of phosphorus, calcium, potassium, and all the other elements, gives increased resistance. Balance is the important thing. "Plant disease depends on a preceding, exactly defined, specific, mineral deficiency . . . plant diseases probably increase parallel with the soil deterioration and have their origin in the deficient or unbalanced plant nutrition" (p. 152, *Fertilizers, Crop Quality and Economy,* 1974).

10. *Other soil conditions.* We covered the vital importance of well-aerated soil in Chapter 3. Anaerobic (non-aerated) soil and high toxin levels are certain to put stresses on plants, kill or inhibit beneficial soil organisms, and lead to disease and pest attack. Proper soil moisture level, pH, and temperature are also important: too high or too low creates stresses.

Grasshoppers are a problem where dry soil conditions exist. Corn borers attack where soil has poor structure, low humus, and low pH and calcium (or normal pH and high magnesium). Cutworms occur in acidic, low calcium soil. Wireworms like tight, wet soil (often because of high magnesium). Japanese beetles tend to occur where organic matter is not decaying aerobically. Even slight soil compaction and restriction of drainage leads to severe infection of *Fusarium* root rot and other diseases in peas in eastern Oregon and Washington. Beneficial soil fungi that fight the disease pathogens are being studied by USDA researchers. During the summer of 1984, the Wisconsin Cooperative Pest Survey Bulletin of June 29, July 27, and August 17, stated that armyworms and corn stalk borers were worse in no-till or minimum-till fields in southwestern Wisconsin.

How to foster natural resistance. What can you do in your fields to promote natural plant resistance to diseases and pests? The answers have been scattered throughout the preceding pages, but let's collect them together and summarize.

1. The overall thing to strive for is a vigorous, healthy, stress-free plant. In reality, there will always be some stresses, but try to reduce those you can control—primarily factors of the soil.

2. The soil should have good structure (tilth). This gives good aeration and drainage. An adequate humus level is essential for good structure. Tillage methods should be appropriate for your climate, soil type, and crops. If soil conditions are not ideal, rotation of crops can give good protection from certain pests, especially those that do not migrate far and those that mainly live in the soil, such as nematodes, wireworms, and gall gnats, as well as many fungal diseases.

3. Soil fertility should be high and balanced. Frequent soil tests should be run, and proper corrective and/or maintenance fertilization should be done. An adequate humus level, regular recycling of organic matter, and a healthy population of soil organisms are the ideal ways of achieving this.

High application of synthetic commercial fertilizers may increase pest or disease vulnerability.

4. High quality seed and crop varieties that are adapted to your climate should be grown.

5. Any materials applied to the soil should be free of (or not produce) toxins that may injure soil organisms or destroy humus. Pesticides should only be used as a last resort. Even organic matter can produce toxins in the soil if not applied properly.

Non-toxic pest control methods. But if your soil isn't yet in healthy condition, or if bad weather puts your crops under stress, and if insects, mites, or nematodes do attack your plants—then what should you do? Drench them with poison?

Try to avoid that drastic option until you look at the situation more carefully. Of course, not all "bugs" on your plants are bad ones. There will be many "good guys"—natural enemies of the "bad guys"—such as green lacewings, ladybugs, damsel bugs, tachinid flies, and parasitic wasps, to name a few. Check out a "bug book" from the library and get familiar with your friends. Learn how to tell a helpful damsel bug from a harmful chinch bug. I have seen tiny parasitic wasps eliminate cabbageworms on a few unhealthy cabbage plants in an otherwise healthy garden. The wasps, resembling gnats, swarmed around the cabbages. The females laid eggs in the cabbageworm caterpillars. The wasp larvae ate the blood and internal organs of the cabbageworms, finally emerging to spin white cocoons on the leaves and change into adult wasps.

Even if there are some pests chewing leaves or sucking sap, is the damage they are doing really significant? Most plants can lose up to 30–40% of their leaf area before they are harmed, and vigorous plants can sustain heavier attack and outgrow it. In a study of pea plants, the plants grown on nutrient-deficient soil suffered most from aphid attack, but vigorous, well-nourished plants actually had more aphids on them.

According to pest control experts, if pest damage is below a certain level, called the economic threshold, economic injury level, or critical level of damage, applying a pesticide or any other control will not pay. This varies depending on the value of the crop and the cost of the control. However, only short-term economics are considered; any long-term damage to the soil and natural enemies is ignored.

Let's say that pest damage is above the economic threshold or soon will be—what is the best control method? There is no simple answer; it depends on the situation—the crop, the pest, the weather, soil conditions,

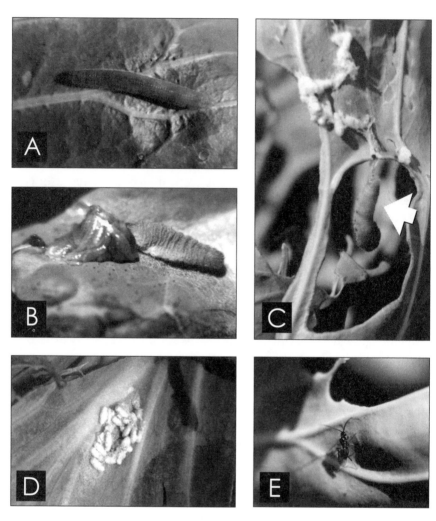

Macrocentrus, *a tiny (¹/₈ inch) wasp parasite of the cabbageworm (European cabbage butterfly). A. normal-looking cabbageworm; B. cabbageworm with many wasp larvae squeezed out; C. dying cabbageworm (see arrow) with wasp cocoons above it on leaf; D. cluster of wasp cocoons; E. adult wasp searching for new cabbageworms in which to lay eggs.*

and whether there are natural enemies. Some careful field scouting and detective work are needed.

   If the soil and weather conditions are favorable for vigorous plant growth, and if a survey reveals plentiful natural enemies (predators, parasites, or diseases), let nature take its course (while keeping a careful watch on the situation). The pests should be controlled by their natural enemies, within a couple weeks usually. If natural enemies are few or not doing their job, you can buy and release natural enemies to help nature along. This should be done about two weeks before anticipated serious pest damage; however, this is usually expensive.

You can buy predators such as adult ladybugs (release ½ to 3 gallons per acre) or green lacewing eggs (10,000 to 50,000 per acre); these are good for small pests such as aphids, leafhoppers, scales, mealybugs, and mites. Three species of predatory mites (1,000 to 10,000 per acre) are available to kill plant-feeding mites on beans, roses, cucumbers, squash, peaches, strawberries, grapes, apples, and citrus. Praying mantis egg cases (10 to 100 per acre) hatch into young mantids that eat aphids, small caterpillars, etc., and then grow into large adult mantids that consume grasshoppers and other large pests. And you can put up birdhouses to encourage wrens, bluebirds, swallows, etc.

Parasitic wasps called *Trichogramma* can be bought in egg form; they parasitize eggs of corn earworms, cutworms, pink bollworms, leafworms, armyworms, codling moths, cabbage loopers, oriental fruit moths, peach tree borers, and other caterpillars. Parasitic wasps for alfalfa weevils, Mexican bean beetles, and certain other pests are being developed. Parasitic nematode worms can be sprayed or scattered in a cedar chip medium for control of corn rootworms, corn borers, cutworms, wireworms, black vine weevils, cabbage root maggots, and carpenter worms.

A caterpillar bacterial disease called *Bacillus thuringiensis* (Bt) can be sprayed, and a milky spore disease (another bacterial disease) is used for white grubs and the Japanese beetle. A virus disease to control the codling moth is available. When ordering natural enemies, state the pest(s) you want to control, since there may be different strains or species for different pests.

Collecting some of the insects that are damaging your crops, grinding them up in a blender with a quart of water, and (after straining) diluting with water and spraying on the plants has been found to effectively kill the pest species. Entomologist F.R. Lawson believes the method may work by spreading an insect disease virus, or possibly by a repellent odor. About a half-pound of insects will treat around 75 acres if approximately 4 oz. of the blender liquid is diluted with 1500 gallons of water.

If you can't wait a couple weeks for natural enemies to do their job, there are other ways of killing pests that do not pollute the environment with highly toxic chemicals. For fruit trees or other trees, the use of a dormant oil spray in late winter or early spring, or a summer oil spray, can control (by suffocation) scales, and aphids. Ordinary sugar, worked into the topsoil will not only kill nematodes directly by drying them out, but also stimulates the growth of beneficial soil microbes. Spraying molasses or other sugar solutions will make plants unpalatable or will suffocate small pests. Gritty powders, mainly diatomaceous earth, are a non-toxic way to kill insects. When dusted on plants, insects or mites pick up the grit and die by drying out when their protective cuticle is scratched. Dusting an animal is a good cure for lice or fleas, and diatomaceous earth given orally (mixed with food) is an excellent de-wormer (about a handful per large animal per day for a week).

Insect traps are being used successfully in certain situations. Light and electrocution traps attract night-flying moths, such as corn borers, corn earworms, cutworms, and armyworms. Pheromone (sex-attractant odors) traps are used to help control fruit flies and Japanese beetles. Or the pheromones are sprayed in the environment to confuse the insects and reduce mating effectiveness. These methods reduce pest numbers somewhat but may not provide effective control.

Natural insecticides (called *botanicals*) made from certain plants are comparatively safe for the environment. They break down quickly. Botanical insecticides sometimes used include pyrethrum (or pyrethrins), rotenone (or derris, toxic to fish; do not use near water), sabadilla, quassia, hellebore, nicotine (from tobacco), and ryania. They would be first choice if your pest problem requires immediate killing of pests.

The long-used insecticides called the *inorganics* include some that are relatively safe ecologically (sulfur, lime-sulfur), but others that are very toxic and which should be avoided (arsenic and mercury compounds). Copper sulfate is an effective fungicide and supplies the trace element copper, but can kill soil life in large amounts.

The least desirable insecticides are the ones most used today, the *synthetic organics*. They have been widely used since World War II and have led to serious ecological problems. Some are slow to break down, while others are rapidly degraded. Over 40 have already been banned or restricted because of their pollution or cancer-causing potential.

Many pesticides are very toxic to honeybees (including carbaryl [Sevin]), parathion, diazinon, ethion, cygon, methomyl, Rogor, Imidan, Dibrom, azinphosmethyl, Furadan, Penncap, Spectracide, Dursban, malathion and phosdrin). Pesticides that are relatively safe for honeybees include methoxychlor, Fundal, most miticides (acaricides), most fungicides, and most herbicides. Bees are more valuable to us for their pollination services than for the honey they produce. Billions of dollars worth of fruit, vegetable, and forage crops depend on insect pollinators.

If you must use a synthetic organic insecticide to save a crop, go ahead, but if possible reduce the dosage. The use of a surfactant (wetting agent) in the spray may allow you to reduce the amount of pesticide considerably. Always follow label precautions and instructions, and NEVER increase the dosage on the misguided philosophy that "if so much is good, twice as much is better."

Animals too. Just as good health and vigor protect plants from their pests and diseases, so also are animals (and humans) protected from parasites and diseases. Are infectious diseases caused by "germs"? Well, yes but . . . we are surrounded by disease germs daily, but as long as we are in good health—get plenty of sleep and eat a good diet—we don't get sick. Usually, it is only when an animal is under stress—in poor health—that disease pathogens can get a foothold.

Healthy animals have various defenses against parasites and diseases, including antibodies and white blood cells. It is well known among animal breeders and geneticists that the offspring of certain crossings are resistant to insects and diseases (J. Blakley & D.H. Bade, *The Science of Animal Husbandry*, 1976, p. 129). Organic farmers often report that their livestock are not bothered by flies. Veterinarian Dr. John Whittaker, writing in *Acres U.S.A.* (Dec. 1975, March & April 1976), states that B vitamins, vitamin C, and other nutritional factors play an important role in protecting animals from parasites and diseases; for example, an imbalance of dietary calcium and phosphorus or a magnesium deficiency increase parasitic worm infestations, while a high carbohydrate diet increases the infection of *Balantidium*, a protozoan intestinal parasite. He notes that too much soluble nitrogen (non-protein nitrogen) or urea in feed causes high blood urea or ammonia levels, leading to reduced resistance to bacterial infections. Resistance to parasites and diseases can also be lowered by vaccinations and antibiotics (these can kill rumen microbes, leading to toxic mold infections), worming medicines, moldy feeds (through mold-produced toxins, including aflatoxin), and stresses (weather, noise, moving, and diet changes).

Nature's way—Weeds. By working with natural systems, effective control of pests and weeds can be obtained. The basic cause of weeds is out-of-balance soil. Weeds are undesirable plants (carryover corn in a soybean field could be called a weed) See the table of indicator weeds earlier in this chapter.

Some weeds are encouraged by the use of herbicides and wrong fertilizer materials (which can throw the soil out of balance and kill beneficial organisms), including foxtail (helped by the high magnesium of dolomitic limestone, which may tighten the soil) and fall panicum (allowed to grow when herbicide treatment removes foxtail). One of the most common causes of weeds is high potassium soil.

Most weeds actually do not grow well in good, healthy soil and in competition with vigorous, healthy crop plants (*Weed Control*, 3rd edition, p. 93). I have seen sick, insect-riddled weeds with hollow stems and rotting roots growing next to healthy corn and alfalfa in fields that have been under a corrective fertilization program for only seven months. A few weeds will grow in good soil, but are seldom a serious problem; two weeds that indicate good soil are Lamb's quarters and redroot (rough) pigweed; in some parts of the world these "weeds" are a valuable forage crop. Certain plants actually exude toxic substances from their leaves and roots, usually while in the seedling stage, which inhibit the growth of, or even kill, neighboring plants. This phenomenon is called *allelopathy*. For example, quackgrass and velvetleaf have been found to inhibit alfalfa seedlings, and oats inhibit growth of the weed charlock up to 38% (*Weed Control*, 3rd edition, p. 103–107). There is some interest among researchers in using allelopathic crop plants or their toxins as natural herbicides (*Crops & Soils Magazine*, March & April 1980; *Agronomy Journal*, Vol. 88, No.6).

Non-toxic weed control. We have seen that most weeds grow best in out-of-balance soil with poor structure. So obviously, the number one priority in solving a weed problem is to correct whatever is wrong with the soil. The three soil goals to work toward are (1) a loose, crumbly soil structure; (2) a good balance of soil nutrients and desirable pH; and (3) abundant beneficial soil organisms. A soil test can spot nutrient imbalances, but a shovel is your most useful diagnostic tool. By simply digging several holes in your field you can determine the general tilth (by ease of digging), spot waterlogging, a hardpan, and anaerobic soil (by how far down small feeder roots extend; also, do buried manure and plant residues decay rapidly?).

But until soil problems are corrected you may have some weed problems to contend with. As with pests, try to avoid use of synthetic chemical herbicides. If they are needed, using a surfactant (wetting agent)

No herbicide was used on this corn field.

may allow you to reduce dosage, sometimes by one-half. Here are several natural or non-toxic methods:

1. *Mechanical control.* Timely cultivation with shovel, sweep, rotary, tine or disk cultivators, rotary hoes, and spike-tooth and spring-tooth harrows is an effective weed control method. Get the weeds as early as possible, before their roots go deep. But be careful not to prune or disturb crop roots. It may take two or three treatments before early weeds are effectively controlled. Cultivation is also beneficial in loosening and aerating the soil, and in dry weather, the "dust mulch" it leaves discourages weed seed germination. Hilling of soybeans and especially corn not only buries weeds but also stimulates corn prop root growth. Frequent mowing of weeds will weaken them by depriving them of food from the leaves. This is useful for already established weeds. If at all possible, prevent weeds from going to seed.

In gardens, orchards and some high-value crops, it is practical to hand-pull or hoe weeds. Hand-weeding can also be used to get rid of small patches of very noxious weeds, before they become widespread.

2. *Seedbed preparation.* For row crops, if you prepare an ideal seedbed in the row but leave the soil rough and cloddy between the rows, the crop should sprout first and get a head start on weeds. Or you can cultivate and delay planting, allow the weeds to sprout, and then till them out as you prepare the seedbed. Use of a ridge-planting system along with cultivating and mowing can effectively control weeds. Vigorous crop plants will shade out weeds and are better able to withstand weed competition. Weeds that appear later in the season decrease yields only slightly.

3. *Crop rotation.* Growing the same crop continuously (monoculture) usually leads to infestation by certain weed species. But growing a series of different crop species (rotation) can not only improve soil conditions and reduce fertilizer needs, but also reduce weed pressure by "keeping weeds off-balance," since each species of weed grows best under certain soil conditions.

When planning a crop rotation, you need to consider overall economics and profitability, equipment needed, crops or varieties well-suited to your soils and climate, and especially a series of crops that will improve your soil structure, beneficial organisms and fertility. Grow a nitrogen-fixing legume just before crops that are high nitrogen consumers, such as corn. Forages and fine-rooted grasses are great for improving soil structure.

A shorter rotation often works better than keeping the same crop for 3 or 4 years. Some rotations that work well in cooler climates are:

a. corn—soybeans—oats/alfalfa—alfalfa—alfalfa

b. corn—corn—oats/alfalfa—alfalfa—alfalfa—alfalfa

c. corn—fall cover crop—corn—fall cover crop—soybeans

d. corn—small grain—oats/alfalfa—alfalfa—alfalfa—alfalfa

4. *Cover crops.* Planting a leafy or dense cover crop will shade out most weeds (also called a smother crop). Small grains (wheat, barley, rye) and buckwheat work well and can also later be turned under or mowed as green manure. Forages such as alfalfa or clover are excellent and help build the soil. Grasses and wood-chip mulches work well in orchards. Planting the shade crop early will allow it to get a head start on the weeds. Intercropping or interseeding, planting a different crop between the rows of a row crop (such as soybeans between corn), helps control weeds, but the yield of the low-growing crop will be decreased. Other good cover crop species include forage brassicas (turnip, kale), canola, sunflower, and mixtures such as oats/rye, rye/hairy vetch, alfalfa/grass, red clover/annual

ryegrass and yellow sweet clover/rye. Although they are not cover crops, corn, buckwheat, or soybeans can shade out non-grassy annual weeds. A cover crop can also be grown as a separate crop in a rotation, typically being planted in late summer or fall and plowed under next spring. If a crop species is grown in a climate for which it is not well adapted, it will have to struggle and will not compete well against weeds. Planting varieties that are adapted to your area is best.

5. *Allelopathy*. One reason some of the above rotations or cover crops control weeds is not only by shading them out, but also allelopathy, the release of toxic substances (phytotoxins) that kill or inhibit other species of plants. The phytotoxins can be released from the roots of a growing plant or can be produced when plant residues decompose in or on the soil, sometimes being released by certain soil microbes.

Both weeds and some crop plants can release phytotoxins. That is one reason it is good to get a crop off to a head start. In fact, phytotoxins from plant residues can injure the next crop in a rotation sequence. In a University of Nebraska study, stubble-mulched wheat residues reduced corn germination by 85% in years of normal or above-normal precipitation (this suggests improper, anaerobic decomposition to me). In descending order, alfalfa, birdsfoot trefoil, ladino clover, red clover, reed canary grass, brome grass, timothy, and orchard grass residues were found to release phytotoxins. Rye, fescue, sunflowers, oats, and sorghum have been tested and found to control weeds, some over twice as powerfully as 2,4-D herbicide. Some farmers and orchardists can save $50 or more per acre using allelopathy instead of herbicides.

6. *Natural enemies*. Some weeds have been successfully controlled by turning loose insects that eat them. Examples include a species of leaf beetle to control Klamath weed (St. John's wort), flea beetles to control purple loosestrife and alligator weed, a moth to control three European ragweeds, a moth and a flea beetle to control tansy ragwort, a moth to control purple nutsedge, seed-eating weevils to control puncturevine, and a weevil to control musk thistle.

Naturally-occurring fungus diseases are being developed to control certain weeds in soybeans, corn, potatoes, rice, tobacco, and citrus. Some are already available commercially, and they are sometimes called "mycoherbicides." Weeds controllable by mycoherbicides include velvetleaf (buttonweed), sicklepod, leafy spurge, prickly sida, northern jointvetch, and citrus milkweed, some of which are resistant to herbicides. The natural control is thus more effective and cheaper than herbicides.

Grazing animals (sheep, goats, cattle, horses, geese, and chickens) have been used to control weeds in orchards and truck farms.

7. *Other methods.* Some rather unusual weed control methods are sometimes used. First is burning, or flame cultivation. A propane flame quickly passing by an early crop will immediately kill or injure weed seedlings (especially if they are under two inches high), but not seriously harm a vigorous foot-high crop such as corn (just a little singeing, which will soon be outgrown). Burning is usually used on in-the-row weeds that regular cultivation doesn't reach. It works best on hot, dry days for grass weeds. Burning may also kill certain insect pests, including potato beetle larvae.

The flame cultivation rig has two or more propane burners per row, aimed downward at an angle. With propane costing only several dollars per acre, burning can be an effective weed control tool.

A somewhat similar heat-source weed control method is steaming, in which propane heats steam to 800°F and emits it through a several-foot moveable arm. The tractor-towed rig has given good weed control in orchards and vineyards, with weed ground-cover reduced from 85% to 2% after three days in tests. After 30 days, regrowth was about 30%.

The use of microwave pulses to "cook" weeds has been experimented with, as have lasers.

A recently developed weed control method is electrocution, in which a 15,000–20,000 volt current, running between copper electrodes on the rig and the ground, zaps weeds that are taller than the crop. The current kills plants by vaporizing their tissue sap. Some weeds are more susceptible than others, with about a 40% kill rate. A good contact between the electrode and weed is necessary. The high price of the equipment makes this method economical only for larger acreages.

Last resort. Let's face it, weeds are a formidable foe. Some years you just can't knock them down, or keep ahead of them. Perhaps a long spell of wet weather prevents cultivation or holds the crop back . . . or more likely, you just didn't have time to use the ecologically desirable non-toxic methods covered above. If the weeds are coming on "fast and furious," you may have to resort to toxic "rescue chemistry," namely herbicides (assuming you aren't organic).

Fortunately, most herbicides aren't as dangerous in the environment as are most insecticides and fungicides. Many ecologically-oriented farmers have been able to achieve good weed control by using commercial herbicides at lower rates than what the label stipulates, sometimes as low as half-rate (although the chemical company does not guarantee the product if it is not used according to their instructions).

The effectiveness of most herbicides can be increased if a surfactant (wetting agent) is mixed with them. You can purchase synthetic surfactants, or use a natural one, such as humic acid. Adding a liquid nitrogen fertilizer to the herbicide tank can increase the effectiveness of some herbicides.

Another way to reduce the amount of herbicide in the environment is to spot-spray the worst areas or band-spray in the row while cultivating between rows. Also, if weeds are taller than the crop, you can apply herbicide with wick and wiper equipment.

Corn stalk rot—why? To close this chapter on what really causes diseases, weeds, and pests, let's look at one specific disease problem which is increasingly common—stalk rot of corn. It is easy to drive around in the corn belt in autumn and see one field after another with stalks broken and lying at crazy angles—a headache to harvest and a terrific loss of yield, from 3 to 18%, according to M.E. Michaelson (*Phytopathology*, Vol. 64, p. 499).

The experts say stalk rot and its companion, root rot, are "caused" by fungi—*Diplodia, Gibberella, Macrophomina,* and *Fusarium* to be exact. But is fungal invasion of roots and then the stalk the real cause? The fungi happen to be common in the soil. Not all plants and not all fields are affected; the disease is worse in some years. What conditions allow the fungi to invade? Let's put together a jigsaw puzzle picture of what really causes corn stalk rot by quoting relevant passages from scientific research reports, available in most large libraries. I have added some translations into everyday language in brackets and emphasized key words by adding **boldface** lettering.

1. "Corn plants . . . approaching maturity are ordinarily attacked rather than young, **physiologically active** [actively growing and metabolizing] plants" *(Phytopathology,* Vol. 55, p. 623).

2. "The spread of D. [*Diplodia*] *maydis* in the corn plant is influenced by the **physiological state** [state of health] of the invaded cells" *(Phytopathology,* Vol. 56, p. 34).

3. "Histological observations [examination of tissues under the microscope] indicated that the spread of D. [*Diplodia*] *zeae* was **limited to dead cells.** Spread of *Diplodia* was **inhibited by living cells**" *(Phytopathology,* Vol. 53, p. 1100).

4. "Pappelis correlated resistance to stalk rot from D. [*Diplodia*] *zeae* with the density of pith tissue in the stalk. Pith tissue of high density had

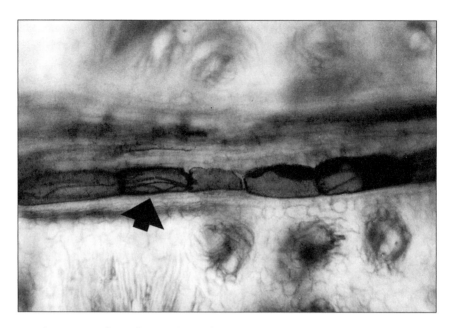

Invading corn stalk rot fungi in base of stalk. Magnified 100 times by microscope.

a high percentage of living cells, whereas dry fluffy pith of low density was composed mostly of dead cells" (*Phytopathology*, Vol. 51, p. 376–377).

5. ". . . root and stalk rot is preceded by a loss of cell turgor [internal "water pressure;" correlated with health of cell] and a **reduction in sugar levels** in the pith, whereas resistance to stalk rot is associated with the maintenance of succulent stalks and **high sugar levels** until after physiological maturity" (*Canadian Journal of Botany*, Vol. 43, p. 1277).

6. "Resistance appears to depend upon the maintenance of a certain degree of **physiological vigor** [good health]. . . Conversely, susceptibility is due to a **lack of vigor** as the plant matures. Physiological vigor depends on a steady respiratory rate supported by a continuous supply of carbohydrate reserve, always assuming an **adequate and balanced level of nutrients**. When the vigor of a plant drops below a certain level because of **stress conditions** under which it is grown, it becomes susceptible to invasion by certain saprophytes [plants living off of dead organic matter] and weak parasites" (*Canadian Journal of Botany*, Vol. 44, p. 451).

7. ". . . **environmental stresses** reduce photosynthetic [food making] capabilities of host plants" (*Phytopathology*, Vol. 70, p. 534).

Weeds compete with the crop for limited water.

8. "... root and lower stalk tissues are decayed by several microorganisms as the tissues lose their **metabolically dependent defense system** . . ." (*Plant Disease*, Vol. 64, p. 534).

9. "The roots were the site of initial infection, and the pathogen progressed into the stalk by way of infected roots" (*Phytopathology*, Vol. 51, p. 382).

10. "... the greater the infestation of the European corn borer, the greater the amount of stalk rot. The holes made by the insect provide easy avenues of entrance for fungi into stalks, shanks, and ears" (*Phytopathology*, Vol. 50, p. 284–289).

11. "An **antifungal** phenolic compound [a natural disease-fighting chemical] has been isolated from corn roots and stalks" (*Canadian Journal of Plant Science*, Vol. 44, p. 451)."... extracts from young corn stalks inhibit growth of two stalk-rotting organisms, *Fusarium moniliforme* Sheld. and *Gibberella zeae* (Schw.) Petch. Loomis *et al.*, found **resistance factors** in corn which were effective against the European corn borer" (*Nature*, Vol. 183, p. 341). "The occurrence of a cyclic hydroxamate . . . (DIMBOA), in maize [corn] was first reported in 1959 . . . Beck & Smissman . . . found that it inhibited insects, fungi and bacteria" (*Physiological Plant Pathology*, Vol. 1, p. 516). "It is suggested that this compound is equally able to play

Weedy corn field.

a role in the resistance of completely developed plants of maize [corn] to injury of borer larvae . . . and of other fungi, like *Fusarium* responsible for stalk rots" (*Annuales des Epiphyties*, Vol. 19, p. 94).

12. "A high nitrogen to potassium ratio greatly increased the severity of internal rot, the amount of stalk breakage, and premature dying. One method to reduce the severity of basal stalk rot or nodal rot may be through the prevention of early death of nodal cells by controlling soil fertility . . ." also "... **unbalanced soil fertility always was conducive to stalk rot of corn** . . ." (*Phytopathology*, Vol. 56, p. 850, 852).

13. "An increase in the concentration of phosphorus . . . reduced the development of both stalk- and root-rot of corn . . . the mineral balance most consistent with good growth of corn and inhibition of stalk- and root-rot is a medium level of nitrogen and potassium and a high level of phosphorus" (*Phytopathology*, Vol. 50, p. 212, 214).

14. "Potassium chloride [muriate of potash], applied at . . . 150 lb/acre potassium, decreased natural root necrosis [decay] and delayed death of corn plants, but **increased stalk rot** when the stalks were inoculated with *Diplodia maydis* . . . rates of [potassium chloride] of 101 kg/ha (90 lb/acre) of chloride caused a depression in plant growth, delayed maturity, and decreased yields" (*Agronomy Journal*, Vol. 59, p. 332, 499–500). "Ammonium chloride treatments were included on selected plots . . . to determine whether or not reductions in lodging . . . were due to chloride or potassium in the KCl [potassium chloride]. It is concluded that the reduction in lodging was due to potassium and not chloride" (*Agronomy Journal*, Vol. 68, p. 425).

15.   "Rate of photosynthesis in each plant is affected by soil microenvironmental [small-scale] factors, leaf damage, and plant competition; but larger macroenvironmental [large-scale] **stresses** such as cloudiness, drought, **fertilizer imbalance, poor soil waterholding capacity,** and leaf disease epidemics probably are more important" (*Phytopathology*, Vol. 70, p. 535).

Summary. In case you got lost, let's summarize. Corn root and stalk rots develop in plants that are not vigorous, with a low sugar content (low rate of photosynthesis), and dying pith cells. Stresses from such things as drought, cloudy weather, high stand populations, leaf damage, and insect attack allow the fungi to get started. But I would trace the real cause back to the soil: **nutrient imbalance** and **poor soil structure.** Healthy plants produce antifungal chemicals, but those under stress apparently do not.

Now that we have seen some of the main causes for our problems, let's look at ways to solve them, to rescue sick soil and get started on the road back to healthy soil.

Chapter 17

RESTORING SICK SOIL

P erhaps you are convinced that natural farming methods are the way to go, but you don't know how to switch over. Or you are worried about lower yields or insect infestations if you don't use the methods that you and your neighbors have been using. These doubts and questions are understandable and common.

You know things are bad and getting worse, and you are looking for a way out of the vicious circle you are in—higher expenses and lower income—harder soil and lower crop quality—more weeds and bugs—sick animals. These are symptoms of problems. Problems that trace back to sick soil.

But they are problems that can be overcome. You need to (a) recognize the problem, (b) find the cause or causes, and (c) eliminate or overcome or prevent the cause(s).

Most sick agricultural soil wasn't that way before it was tilled. Virgin soil is usually pretty fertile and has good tilth. It is from unwise use (or abuse) by man that soil has deteriorated. It probably took years, even decades, for the damage to be done. It may take a number of years to make much progress in some cases, while in others amazing changes can be made in one year. So, to correct problems in the soil, the basic approach to use is to (a) stop doing wrong things, and (b) start doing right things.

Restoring sick soil. Let's look at some practical methods for restoring soil fertility and good tilth to abused soil. We will first look at methods that should work on most types of soils and then later we will cover methods of dealing with some problem soils, such as acid soils, alkaline and dryland soils, waterlogged soils, and soils polluted by toxic substances.

Problem analysis. The first thing a doctor does before treating a sick patient is diagnose the illness—to use his trained eye, knowledge, and perhaps some lab tests to find out what's wrong.

So before rushing out blindly to correct your sick soil, you had better have a fairly good idea of what is wrong with it, so you can do a more effective job and keep from making costly mistakes.

If you feel inadequate to diagnose your soil's ills, you may want to hire a professional consultant, although it can be risky if he also sells a product. His product may help, but you might make faster progress correcting sick soil if he can recommend a variety of materials, depending on need. Actually, for many if not most sick soils, if you have a little background knowledge of how soil and soil life function to nourish a plant, and if you have an observant eye and a lot of common sense, you can probably correct 90% of your soil's problems yourself.

Call a spade a spade. The first and most important "scientific instrument" you should use to test your soil is a shovel! A long, narrow spade is best. Walk around in your fields and observe your soil carefully. Does it feel spongy when you walk on it, or does it more closely resemble a parking lot? Does it crust when dry or stick to your shoes when wet?

When you attempt to thrust in the spade, does it sink in at least six or eight inches with little effort, or do you have to jump on the spade to get it in two inches? When you pull the spade handle back, does the soil crumble into loose clumps or does it fracture into hard clods? See any earthworms?

Try digging deeper than the "plow layer." Do you hit a dense layer—a hardpan? Dig down through it and see if the soil is looser below it. In wet weather, the soil above a hardpan may be waterlogged, while below it is bone-dry.

Are there crop residues such as corn stalks several inches down that are a year or more old and still not decayed to humus? If so, maybe the soil is anaerobic or the beneficial soil organisms have been killed by toxins.

Take a lump of topsoil in your hand. Is it easy to crumble? When you squeeze some damp soil in your fingers, does it become a compact wad of clay? Smell your soil. Does the topsoil have that rich, pleasant odor you associate with humus or a fertile garden soil? If so, beneficial actinomycetes are alive and well (see Chapter 3). If soil (especially subsoil) has a foul, putrid odor, anaerobic fermentation is occurring, probably with toxins being produced.

Compare these observations in soil taken from good fields and problem fields, or problem spots in a field. You can tell a great deal about your soil when you get off your tractor and "get down to earth."

If you have read Chapters 3 and 6, you can probably analyze some of the most common soil problems: low humus, and sterile, tight, compacted, anaerobic soil. But there's more.

What plants tell. Dig up the root system of a plant, such as corn or soybeans, or even pigweed. How far down are there a lot of fine feeder roots (taproots may extend deeper)? This shows the depth of your soil's aerobic (aerated) zone. Do your soybean, alfalfa, clover, snapbean, or other legume roots have the nitrogen-fixing root nodules that they are supposed to? Nitrogen-fixing bacteria need well-aerated soil to form nodules. If there are nodules, cut some open. Are they pink inside or a sickly yellow or greenish? Only pink nodules are doing their job. Soil must be free of toxins in order for nodules to grow and function properly.

If your soil is plagued by weeds, check the box "Indicator Weeds" in Chapter 16 to see if they indicate a particular soil problem. But remember that a few kinds of weeds indicate good soil.

Are the plants growing in your soil healthy looking? Have yields or quality been poor in past years? A lot of bugs or diseases? These are more indications of sick soil—perhaps anaerobic or toxic.

Deficiencies of mineral elements can cause visible symptoms in the leaves of plants, but their diagnostic value is limited. They usually show up only when the deficiency is severe, and by the time you see them, it may be too late to correct the problem that year. Also, there may be more than one deficiency at once, with diagnosis being difficult. Nevertheless, the accompanying box lists the most common mineral deficiency symptoms that apply to most crops (a few crops may react differently).

The "deficiency" of an element may be due to different causes. There may be a true deficiency or lack of the element in the soil. Or there may be plenty of the element, but it simply is not available to the plant in sufficient amounts. This can be caused by high or low pH (see diagram in Chapter 12, Availability of Elements) or by an imbalance (too much or too little) of another element (see table in Chapter 10, Nutrient Interactions). And then there can be plenty of an available element, but if the plant is stressed or sick, the roots may not be able to absorb it. Stresses can come

General Mineral Deficiency Symptoms of Plants

(compiled from various sources)

- **nitrogen**—slow growth of top and roots; leaves turn yellow-green when young to orange, red or purple when mature (veins may turn purple), beginning from bottom leaves to top; reduction in number of flowers and yield of grain or fruit; slow growth and delay of opening of buds.

- **phosphorus**—similar to nitrogen deficiency but leaf color either dull bluish-green with tints of purple instead of yellow or red, OR leaves dull bronze colored with purple or brown spots (leaf edges often brown, especially in potatoes).

- **potassium**—if mild deficiency, stems thin; if severe, stems stunted or dead; leaves usually dull bluish-green, often yellow streaks between veins, followed by browning of tips and edges, and development of brown spots near edges; leaves rolled (begins at lower leaves); poor root growth; poor development of flowers and grain or fruit.

- **calcium**—begins at upper leaves: leaves very distorted and curled at edges; edges appear ragged and leaves may have thin yellow bands or be brown, then die; roots poorly developed.

- **magnesium**—leaves turn yellow (sometimes between veins, sometimes in spots or streaks), then turn brown and die (starting at bottom of plant); grain or fruits poorly developed.

- **manganese**—similar to magnesium, but starts at top of plant.

- **sulfur**—reduction of growth; curling down of leaves at tips; similar to nitrogen deficiency except that lower leaves are not lost.

- **boron**—begins at upper leaves: leaves become light green (lighter at base, veins turn purple, leaves may have faint streaks and yellowing, then turn orange) and stop growing, roll down at tip, with tissue breakdown at base; poor growth of tops and roots; flower buds drop off; no grain or fruit.

- **iron**—severe yellowing of newer leaves (either spotted or total); more often visible in trees than yearly crops.
- **zinc**—mottled spots on leaves, first yellowish, then yellow or purple-red (appears late in summer, then leaves drop early), leaves become crinkled and small.
- **molybdenum**—similar to nitrogen deficiency, since molybdenum is necessary for nitrogen-fixing bacteria; leaves becoming yellow between veins, first on mature leaves, then to younger; young leaves may become severely twisted and eventually die.

Potassium-deficient alfalfa

from drought, flooding, anaerobic soil, cloudy and cool weather, and toxins in the soil.

So again, plant deficiency symptoms have their limitations in diagnosing soil problems. For similar reasons, the more sophisticated plant tissue testing is also limited in usefulness. It tells you—very accurately—how much of each element is present in the plant, but cannot tell what is wrong in the soil.

Soil tests. Another tool to help diagnose soil problems is soil testing. However, many farmers and advice-givers assume that soil tests are super-accurate and that you can tell exactly how many pounds per acre of fertilizer you need to add this year.

The truth is, there are a whole lot of variables to consider. First, there are many different methods of testing soil and different philosophies of making recommendations based on test results. You can literally send identical samples to two (or more) labs and get two (or more) completely different sets of figures and recommendations back. The numbers aren't wrong; it's just that different ways of measuring something were used. Also, many labs recommend too high amounts of some fertilizers, just to "be on the safe side."

As we mentioned in Chapter 9, most labs use methods that test the soil under artificial (dried, finely ground) rather than field conditions. They generally determine the total amounts of some nutrients rather than what is readily available to plant roots at any one time. A more realistic method uses weak extracting fluids that simulate the nutrient-extracting power of roots. Most labs will run these soluble, readily available tests if requested.

Many farmers have their soil tested once every year or two at the most. But soil is constantly changing, and nutrients that may be unavailable in March or September may be in abundance in June, through microbial action. It may be best to monitor soil fertility at least two or three times a year (spring, summer, and fall) to see what is going on and to spot developing problems. That way corrective measures may be made the same year and a crop saved.

Soil testing kits are available which allow the farmer to quickly get a rough idea of his soil nutrient levels and pH. If you build up a few years of soil testing records and compare results from good and poor fields, you will soon be able to interpret your tests yourself. Many of the professionals tend to recommend too much potassium and too little calcium and phosphorus. See Chapter 9 for more details on soil sampling and testing.

Can't always work miracles. Every field is different, climate varies, and different crops have different requirements. Therefore I cannot tell you exactly what to put on your fields to correct soil problems without knowing the details. In this book I can only give general principles and methods that should work in most cases, on most kinds of soil. But unusual conditions and soil problems do exist which may require the expertise of a reliable consultant. The methods and materials I mention should help when used properly, but I cannot guarantee total success without knowing the situation firsthand. If you know the basic principles of how soil, soil life, and plants function, you should be able to use common sense to decide what to do and to diagnose problems.

General guidelines for soil correction. Following are some general guidelines for correcting abused soil. They should work for most types of soil under moderate climatic conditions, such as would exist in the central and eastern United States. In the dry areas of the West and Southwest (without irrigation), other problems may occur, such as alkaline and saline soils. Guidelines for these and other types of problem soils will be given later in the chapter.

In these General Guidelines, I am assuming that your soil may have such problems as low humus, little soil life, compaction, low or unbalanced fertility, weeds, hardpan, waterlogging, and toxicity from pesticides or wrong fertilizers. So where do you begin, with any or all of those problems? What do you do first?

1. *Analyze the problems.* The first thing to do is to figure out what is wrong. Follow the pointers given just previously in this chapter. Walk in your fields, dig, and observe the symptoms. Make a list. Have your soil tested, especially to spot serious deficiencies or imbalances.

2. *Make plans.* Decide on your short and long-range "plan of attack." If you have relied heavily on chemical pesticides and want to switch to natural "non-chemical" methods, it is best to do so gradually, over anywhere from three to eight years, and to work intensively with smaller fields (up to 30–50 acres) on about 20–25% of your farm. Of course, owned land should have priority over rented land.

If you have livestock or other animals that you raise the feed for, be sure to restore those fields first, and leave cropland that you raise cash crops on for later. Remember to put extra effort into correcting your garden, since you and your family are most important.

If you are rotating crops, use a rotation that will do the most to improve soil and control weeds. Eventually, when the soil is in good condition, you may not even need to rotate (see Chapter 14).

Figure out your budget: how much you will spend on fertilizers and soil amendments, extra fuel for cultivating weeds, and how much you will save by not using pesticides. Allow some leeway for emergency pest control the first couple years.

Get in mind what you are trying to do. Depending on what your exact problems are, you will probably be (a) *improving soil structure* (tilth), loosening compact soil in order to allow air in, which will (b) *increase beneficial soil organisms*, which will (c) *increase humus content* and (d) *improve plant health* and resistance to diseases and pests; while at the same time, (e) *correcting soil nutrient imbalances* and (f) *detoxifying soil*, which will help in all of the above, as well as (g) *reducing weeds*.

Don't expect to do all of that the first year. It will probably take several years to show great improvement, but once you stop doing harmful things and start doing helpful things, rejuvenation will begin, and improvement should be noticed the first year. Keep good records of all that you do: materials applied and amounts, tillage, crops and varieties grown, weather, and results.

3. *Improve soil structure.* There are several additives and methods that can improve soil structure, including calcium, organic matter and commercial products, such as microbes and wetting agents. Some methods and materials are definitely more expensive than others; we will say more about some of them later in this chapter.

A. *Rock fertilizers* can loosen soil. Soft rock phosphate (colloidal phosphate) and high calcium lime do this, plus supplying phosphorus, calcium, and trace elements. These two work best together, at a ratio of two parts lime to one part soft rock phosphate, unless soil is already high in calcium (if applied separately, apply the soft rock phosphate first). Certain other commercial soil conditioner products also help loosen soil and supply minerals, such as humates.

B. *Organic matter*, whether fresh or as humus, will loosen soil. Humus or compost is much preferred, because if too much fresh organic matter (green manures or animal manures, crop residues) is incorporated in tight soil, it may turn anaerobic and release toxins, worsening the problem. However, coarse crop residues (corn stalks, stubble), if partly incorporated and partly exposed, will help air to enter the soil.

Animal manure sources are preferred since they are richer in nitrogen than plant matter, but tests have shown that a mixture of animal manure and plant matter gives better results, and a mixture of manure and rock fertilizers works even better. If animal manures are not available, animal wastes such as tankage, blood meal, fish meal, etc., or high nitrogen plant matter, such as spoiled alfalfa or clover hay, are good (use small amounts at first). Sewage sludge, cannery wastes, old straw, city grass clippings and leaves, and peat are other excellent sources of organic matter.

If the organic matter decomposes aerobically, it will supply most of the soil nutrients needed by plants. But if a lot of fresh organic matter is applied too soon before planting, its decomposition will temporarily tie up nutrients and starve the crop. If the organic matter is low in nitrogen, microbes will temporarily "rob" nitrogen from the soil. Fresh organic matter is best applied in the fall in northern climates, or at least a few weeks before planting in the spring. Compost has the advantage of being already decomposed, but is still rich in nutrients. Compost helps sandy soils the most; fresh organic matter helps tight clay soils more than compost. Compost raises the soil's organic matter content over the long run much more than fresh manure. See Chapter 13 for more details.

C. *Cover crops and sod* are one of the best ways to loosen soil and add organic matter. Perennial grasses, with their multitude of fine roots, penetrate tight soil and add much humus as the roots gradually die. These include such species as smooth brome, orchard grass, fescues, bluegrass and timothy. Legumes (alfalfa, clovers, vetches, etc.) are excellent since they add nitrogen if soil conditions allow root nodules to form. It is best to let the cover crop grow for more than one year. If you do not need the hay, cutting and letting the plant matter lie and decompose adds greatly to soil humus (or do so with the first cutting of alfalfa and harvest the rest).

If the soil is so poor that the above crops will not grow well, just let it go to weeds for one or two years. Deep-rooted annual weeds break up a hardpan, aerate soil, and bring up minerals from the subsoil. Common deep-rooted (with a taproot) weeds include pigweed, goldenrod, ragweed, sunflowers, thistles, nettles, Lamb's quarters, horseweed, wild lettuce, milkweeds, smartweeds, ironweed, dandelion, and wild sweet clover. Shallow-rooted or grassy weeds are not as valuable, but do cover the soil and prevent erosion, plus they can be plowed under as green manure. Weeds should be mowed several times, or at least before they go to seed. Legumes and annual weeds do take a lot of moisture out of the soil, which can be harmful in dry climates. Perennial grasses do not use as much. When any of the above cover crops are plowed under, do so as late as possible in their development but while they are still green. See Chapter 14.

D. *Humates* and humic substances, brown coal, and peat are something like "preserved humus," and often have beneficial effects on soil structure, loosening and aerating soil, as well as stimulating microbe and plant growth.

E. *Seaweed* products, and live *algae* inoculants, can improve soil structure by releasing glue-like substances. A healthy population of *soil bacteria* and other microbes also do the same thing for free. They need food to live on, namely fresh organic matter. Commercial bacterial inoculants are not always successful in establishing beneficial organisms, either because of competition from existing soil microbes, toxic substances, wrong pH, or lack of food (organic matter). The best way to establish beneficial soil microbes is to add a lot of organic matter (animal and/or green manures, crop residues, compost), but not too deeply in the soil (not over 5–6 inches deep). *Enzyme* sprays have also been found to improve soil structure, perhaps by stimulating soil organisms.

Earthworms do a tremendous job in loosening, aerating, and enriching soil. They can break up a hardpan. They do need food (organic matter) and freedom from toxins. They will generally come into an area naturally when conditions are favorable for them (however, they do not like extremely sandy soils).

F. *Wetting agents* or surfactants often do a quick job of making soil more porous and improving drainage. But in other cases results are disappointing, or if soil has little nutrient-holding capacity (sandy soils), nutrients can leach away. Sometimes surfactants can alter soil structure and reduce available living space among soil particles for microorganisms.

G. *Mechanical aeration and subsoiling* are quick ways of loosening tight soil and breaking up a hardpan, but if nothing else is done to improve soil structure, the soil soon settles back to its original condition. They would be good to use in very severe cases *along with* organic matter or other methods mentioned above. Rotary tilling organic matter into the upper several inches is an excellent practice, especially for poorly-drained soil. Injecting manure into the crevices made by a subsoiler has been successful in loosening tight soil.

4. *Correct major pH and nutrient imbalances.* At the same time that you begin improving soil structure, if soil tests or plant symptoms indicate seriously out-of-balance nutrients or much too high or low pH, these should be dealt with immediately.

Ideally for most areas and most crops, pH should be between 6.0 and 6.8, but in good soil with high humus and beneficial organisms, crops can grow well within a larger range of pH, say between 5.5 and 8.0. High humus and beneficial organisms tend to correct pH problems. Also, pH can change greatly in a growing season (in the Midwest, it often drops during the summer), so a single pH test reading shouldn't be taken too seriously. But again, an extremely high or low pH should be corrected.

In average soils, high calcium lime is generally the best material to raise a low pH (acid) since it also supplies calcium, which is usually needed anyway. Dolomitic lime should only be used if the soil is very deficient in magnesium; it is a very hard stone and slow to become available, possibly not for a year or 18 months. The finer the lime is ground, the more quickly it will act. See Chapter 12 for more details.

High pH (alkaline) is seldom a problem except in the West. These problem soils will be covered later.

Unless you know your soil has a serious trace element shortage, don't worry about the minor elements until you have the major elements in balance. The most important are calcium, phosphorus, potassium, and nitrogen. Generally, after these are in balance, there will be fewer trace element problems because humus and soil organisms make them available.

5. *Build up humus.* As we have seen in previous chapters, adequate humus and the beneficial soil organisms that produce it are essential for healthy soil. Most soil organisms need oxygen, so well-drained and aerated soil is required. But humus and soil organisms help make soil loose and well-aerated. So if you are starting at near zero, with "dead" tight soil, how do you start turning things around?

First, reduce or eliminate toxic materials that harm soil life and tighten soil (pesticides, anhydrous ammonia, high salt fertilizers, and excess magnesium on clay soil). Then, use one or more of the methods for improving soil structure previously covered in section 3. Then, since adding fresh organic matter to tight soil can cause more problems, the best way to increase humus is to add compost. Any amount can be added at any time without harm, except that the nitrogen in compost may be too much for small grains if applied just before planting, causing lodging.

If compost is out of the question, fresh organic matter, in moderate amounts at first, should work, if it is kept in the upper several inches. A cover crop and green manure plowed or disked into the upper several inches will also get the ball rolling. Adding rock fertilizers along with the fresh organic matter helps greatly. There are commercial sprays of beneficial microorganisms and/or enzymes that will speed decomposition of plant residues.

After tight soil begins to loosen up and the aerobic zone deepens to 5 or 6 inches, the amount of fresh organic matter can be increased (but not just before planting). Then, several years later, when the soil really is in good condition, you will need less, since soil organisms, air and water will supply most nutrients (soil life breaks down minerals and fixes nitrogen; rain brings nitrogen, sulfur, etc.). Of course some crops require more nutrients than others: corn needs a lot, small grains need little. Still, the soil organisms do need some food (fresh organic matter) to live on, so some should be added yearly. Remember that plant roots, stubble, and stalks add much.

6. *Detoxify soil.* Generally, by the time you have done the above steps, any toxic residues will have already been eliminated. The following are excellent detoxifiers: (a) calcium (lime), (b) soil life (especially bacteria), (c) organic matter (humus, compost), (d) plowed-under green manure (chlorophyll), (e) humates, seaweed, enzymes, etc., and (f) oxygen (keep soil well-aerated, with good tilth).

Most toxic residues will disappear in time, within three to five years.

7. *Control weeds, pests and diseases.* Again, if you have done the above things, your weed, pest and disease problems should already be lessening. Aerated soil with high humus and balanced nutrients should favor vigorous crop growth and discourage most weeds (especially keep the level of available potassium below available phosphorus). Healthy plants will resist pest and disease attack.

What weeds there are can be best controlled by cultivation, in row crops. Cultivation not only kills weeds but performs the valuable job of aerating soil, stimulating root activity and beneficial soil organisms.

If there is a weed, pest, or disease problem that is urgent and cannot be controlled by other methods (see Chapter 16), then the use of toxic herbicides, insecticides, fungicides, etc., can save the crop. If possible, use reduced rates of these dangerous chemicals; adding a surfactant, or wetting agent, will increase effectiveness, allowing you to use less pesticide. Banding pesticides in the row will reduce the amount sprayed.

Guidelines for correcting problem soils. The above General Guidelines should work for most types of soils in moderate climates, but some special treatments may be needed for certain problem soils.

1. *Extremely acid soils; high organic matter soils.* Generally these two go together. In cool climates or where lack of aeration slows or prevents microbial action, organic matter builds up rather than changing into humus. Soils with 20–50% organic matter are called muck soils, while those with over 50% are called peat soils.

In other areas sandy soils can become depleted and acid. Over-use of ammonia, urea, and ammonium fertilizers can lead to too-acid soil, probably with low organic matter.

Of course, one remedy for acid soils is liming. Liming materials contain a carbonate (CO_3^{--}) or hydroxide (OH^-) ion, which combine with hydrogen ions (H^+), which cause most of the acidity. The most commonly used liming materials are ground-up limestone, but there are two main types: high calcium lime (calcium carbonate, $CaCO_3$) and dolomite (calcium magnesium carbonate, $CaCO_3 \cdot MgCO_3$). Dolomite is a hard rock and slow to go to work, plus excess magnesium in some soils can lead to tightening and hardening. High calcium lime thus may be preferred. Papermill sludge is an excellent source of high calcium lime, although it may be hard to spread. See Chapter 12 for more details.

But lime does more than reduce acidity. It loosens compact soil, stimulates microbial action, supplies calcium and some trace elements, and reduces toxins. Soft rock phosphate or basic slag would also help acid and clay soils.

In high organic matter soils, aeration of the soil by tillage and drainage (if waterlogged), speeds up the oxidation of organic matter, which will bring the soil into a better balance. In low organic matter acid soils, adding organic matter will reduce the acidity.

Reducing acidity will generally correct trace element deficiencies and/ or toxicities that tend to accompany acid soils. Adding compost or manure will inoculate "dead" acid soils with soil organisms, even though they may already be high in organic matter.

2. *Alkaline and saline soils.* In the arid western U.S., including irrigated areas, high rates of moisture evaporation concentrates salts of sodium, calcium, potassium, magnesium, and other elements. Also, very close to any ocean, low land can have too many salts. These soils have low availability of phosphorus, potassium, iron, and other trace elements. Soluble salts and/or certain elements in high amounts are toxic to plants. Soil tilth is usually poor and drainage often restricted, resulting in anaerobic conditions and crusting. Alkaline soils have high pH; saline soils have high salt levels.

Alkaline and saline soils must be well-aerated and drained to be corrected. Drainage allows excess salts to leach away. Subsoiling or deep plowing may help in some cases, but soil additives are also necessary. Applications of gypsum (calcium sulfate) or other sulfate-containing fertilizers produce acid to neutralize high-alkaline soil. Iron sulfate is expensive, but would be good if there is an iron deficiency. Even sulfuric acid (battery acid) can be sprayed to reduce alkalinity, but application of pure sulfur requires time-consuming (2–3 months) bacterial action to be effective. If an alkaline soil is low in calcium, a calcium material must be added (gypsum, rock phosphate, superphosphate).

Application of irrigation water after the addition of the above will speed up the action if there is good drainage.

Organic matter, preferably compost, or else sawdust, wood shavings, peat, or cottonseed meal, should definitely be added in the upper layers of the soil, to improve soil structure, reduce toxicities, reduce alkalinity, and inoculate with soil organisms. If you make your own compost, add rock phosphate rather than lime. After soil starts to improve, fresh organic matter can also be added; cover crops and green manures are excellent. Growing a sod crop for a while will improve structure. If soil tests show iron, phosphorus, potassium, or other deficiencies, supplements may be needed until soil improves more.

Grow crops that are salt and alkaline tolerant, such as sugar and garden beets, cotton, barley, asparagus, tomato, carrot, potato, peas, lettuce, alfalfa, sweet clover, wheat, rye, sunflower, certain grasses, etc. One of the best ways to correct saline soil is to grow alfalfa for four or five years. Plant row crops on ridges so salts will leach away from roots. Avoid irrigation with saline water. If ridges are used with irrigation, salts tend to accumulate on the highest part of the ridge during dry periods and can kill germinating seedlings. It is best to plant rows on the sides of the ridge. Keep the soil moisture level high while plants are growing since this dilutes salts.

3. *Prairie and dry soils.* West of the Mississippi in the central U.S. and extending up into the Northwest (eastern Washington and Oregon), there are soils that receive minimal rainfall, are usually slightly alkaline and high in calcium, and (usually) low in humus. These areas are best suited for grazing and wheat or cotton, but with irrigation, other crops can be grown. Unfortunately, excessive and wasteful irrigation practices have seriously depleted the underground water supply in the Great Plains (the Ogallala aquifer). Unirrigated dryland farming should strive to conserve moisture. Crops such as corn and legumes, which require much water (thus irrigation), are not good investments in the long run. It is difficult to raise humus levels above a few percent, but the more organic matter there is, the better the soil will soak up and hold the little rain that falls. Humus will even absorb dew when there is no rainfall. Compost is the best way to add organic matter because it is already decomposed; fresh manure or plant residues may not be able to rot quickly in dry soil. The stubble mulching system (leaving most crop residues, such as wheat straw, on the surface) has been shown to give higher yields and leave more organic matter and nitrogen in the soil than conventional methods.

Leaving the surface cloddy and/or with stubble exposed will aid in accumulating moisture (including snow) and reducing wind erosion. Periodical fallowing, generally with a stubble cover, is often used in dry areas to allow soil to recharge moisture and fertility levels. However, excessive growth of weeds results in moisture loss; cultivation with an undercutter can control them.

Depending on soil tests, fertilization with nitrogen or phosphorus may be necessary. Proper fertilization of crops will actually conserve moisture since healthy plants produce greater yields per unit of water, although they may require slightly more total water.

4. *High rainfall soils.* In the southeastern U.S., along the Atlantic coast, and near the Pacific coast of Washington and Oregon, the soils have been weathered by high rainfall. Subtropical and tropical countries often have the same problems. High rainfall causes severe leaching of certain nutrients (calcium, nitrate), plus erosion. Hot temperatures oxidize organic matter. Consequently, high rainfall soils tend to be acid and low in humus and nutrients.

What these soils need, in general, is organic matter and lime, plus any other nutrients or trace elements in which they may be deficient. These will need to be replenished frequently, because of the constant leaching. Humus will hold and make available a lot of nutrients, but in warm climates, fresh organic matter or compost must be frequently added.

In the meantime. While you are working with your soil, trying to get it back in shape, you have to make a living and raise adequate crops. The best approach is to do as much as your time and budget allow to correct soil problems, and in the meantime, use fertilization and management practices that are the most economical and in tune with nature. In-row supplementary fertilization allows you to grow excellent crops on less than ideal soil, economically.

Soil management. Soil management includes everything you do to the soil: crops grown and their sequence (cropping system), use of fertilizers and pesticides, tillage, irrigation and/or drainage, terracing, etc. Obviously soil management—how you treat the soil—has everything to do with the condition and fertility of your soil 5, 10, and 50 years from now, as well as the yields and quality of your crops, the health of your animals, and your profit/loss figures.

We have already covered a great deal about improving the soil, but now let's say something about cropping systems and tillage.

Cropping systems. There are many different ways of growing crops, but they can be boiled down to two basic approaches: growing a sequence of different crops on the same piece of land (rotation) and growing the same crop year after year (monoculture, continuous cropping). Both approaches are championed by some and criticized by others. There are advantages and disadvantages to both; it depends on your particular situation as to which may be best for you. Either one can be unsuccessful if not done properly. We have previously mentioned them in Chapter 14.

Rotation. The advantages given for crop rotation include its ease of maintaining or building soil organic matter, and supplying much crop nutrition (especially nitrogen) from plowed-under forage crops. Also, rotation helps prevent pest, disease and weed problems by growing different kinds of crops in the same field. Under average soil conditions, rotation can give about 10% higher yields than continuous cropping.

Continuous cropping. The proponents of continuous cropping say that you can match crops with soil type and topography—grow row crops on level land to reduce erosion and keep hilly land in forages or pasture. Or you can specialize and grow the same thing on all your acres, simplifying your machinery roster. Since different crops do best under different fertilizing programs, you can "fine tune" soil to grow top yields of any particular crop.

Disadvantages of monoculture include soil depletion if crops are continuously harvested and if little organic matter is returned to the soil. Row crops and vegetables are especially bad for this. Soil-borne pests and diseases can build up, and certain weeds may increase. With row crops, erosion can be tremendous. Considerable outside fertilizer may be needed.

Which? Whether you rotate or use continuous cropping depends on your needs, soil, and topography. If your soil is depleted and you have no source of much organic matter to recycle (or compost), rotation is the best way to build up soil, with emphasis on forages and/or pasture grasses. Proper recycling of manure and other organic matter can build up soil under continuous cropping. You can do a great deal by growing and incorporating a cover crop (rye is excellent for this).

Some good crops to add nitrogen to the soil are alfalfa, sweetclover, and red and Ladino clovers (soybeans, although they are legumes, add little nitrogen to the soil because they use most of it in seed production). Grasses and legumes provide a slow release of nitrogen when plowed under. To increase soil phosphorus, grow alfalfa or plow under sweetclover. Buckwheat will increase a number of soil minerals. Perennial grasses are excellent for improving soil structure.

White (Ladino) Clover, one species of nitrogen-fixing legume, also builds soil humus.

New seeding of alfalfa with oats as a nurse crop.

After you have built up depleted soil, you may want to consider "fine tuning" some fields to grow top quality forages or high value row crops continuously, with careful monitoring of soil conditions to prevent depletion.

Intercropping. The practice of planting two crops together is not new, and although it has not been popular, interest in intercropping is growing. With a crop such as corn, wide-spaced rows are used and another crop such as soybeans, forages, small grains, or sod is planted between the rows. This has the advantage of excellent weed control, plus the yield of the second crop. See Chapter 14 for details.

Tillage. Tillage is done for three reasons: to change soil structure and prepare a seedbed, to control weeds, and to manage crop residues and organic wastes. Many kinds of tillage methods and machinery have been used over the decades. Some work well in certain types of soil or climate; for example, disk plowing is good for dry climates and light soils, while chisel plowing does not work well for moist soils or sod.

Each type of tillage has its own place and best uses. Moldboard plowing, which has come in for considerable criticism because it inverts the soil and disturbs the soil's ecosystem, still is useful for breaking sod and turning under green manure crops. If the moldboard plow is adjusted so that the slices are stood up on edge rather than being inverted, most of the undesirable results are prevented. This type of moldboard plowing in the fall, followed by freezing and thawing, is good in heavy soils and can break up a hardpan.

In order to do the most good in improving soil structure and humus content, organic matter must be worked into the soil, not left on top. But at the same time if the lower soil layers are tight and not well aerated, fresh organic matter should not be plowed in that deep, but kept in the upper several inches. And of course in some cases, some trash or a thin mulch should be left on top, such as to prevent wind erosion in the plains region.

In some cases or for some purposes, ridge-tilling is excellent. If organic matter is worked into the upper several inches, decomposition to form humus occurs quickly because the ridges provide better aeration. In northern climates, ridged soil warms up fast in the spring if the ridges run east-west. In dry areas, the ridges catch rain or snow. Ridges are commonly used for irrigation and truck crops.

Spring primary and secondary tillage can reduce the earthworm population up to 75% in the upper eight inches, according to studies at the University of Guelph, Ontario, Canada. Although the worms can restore their population levels by fall, summer is the time when they are most needed. Microorganism populations can also be upset. Shallow spring tillage is best, with whatever deeper tillage is needed being saved for the fall.

To till or not to till? The biggest trend these days is to reduce the number of trips across the field because traffic tends to compact soil and waste time and fuel, and because excess tillage causes soil organic matter to be oxidized faster. Also, the threat of erosion is increased. So the big thing being pushed today by the USDA, university extension, and others is no-till, or reduced tillage called by various names: minimum-till, conservation-till, etc. These methods are touted as saving fuel, reducing compaction and erosion, conserving soil moisture, and even building-up soil organic matter. We have covered the pros and cons of reduced tillage in Chapter 14.

Alternatives. Are the ways promoted by the "standard" farm magazines and recommended by the state universities and county extension agents the only way to farm? Are commercial fertilizers, pesticides, and hybrid seed "required" to get decent yields? Are one-fourth of young calves or pigs expected to die from disease and birth defects?

As you probably know from reading the earlier parts of this book, not everything about modern agriculture is necessarily good for the soil, nor produces high quality crops or healthy animals. In previous chapters we pointed out some of the problems caused by pesticides and other practices and gave some ways of improving soil and crops—things that were well-known in the past, but have been shoved aside by the glitter of new technology. Even "the establishment" has finally started to recommend some of these things in recent years, such as reducing the amounts of pesticides and nitrogen fertilizer, split applications and in-row application to conserve fertilizer, rotational and grass grazing, and recycling organic wastes to build up soil organic matter. "Organic farming practices work," says Robert I. Papendick, USDA soil scientist, "that is no longer a disputed fact among knowledgeable people" (*Farm Journal*, March 1982, p. 46). USDA official Paul F. O'Connell wrote, "Much of the public's growing interest in sustainability is a reaction to the sum total of modern agriculture. The public wants an agriculture that will not only be productive and profitable, but that will also conserve resources, protect the environment, and enhance the health and safety of the public" (*Yearbook of Agriculture*, 1991, p. 176). Several state universities are teaching courses and doing research on organic and sustainable methods. The big chemical companies are scrambling to do research on soil microorganisms and non-toxic methods of disease and pest control, mainly to develop their own patentable strains of microbial inoculants or pest-fighters. They are also buying up smaller companies that already have such products. That should tell you something about the trend of the future.

Yes, there are alternatives. Let's look especially at some of the lesser known methods and products, which we briefly mentioned earlier, some of which may become the "traditional agriculture" of the future.

Soil conditioners. Starting with the soil and improving soil structure (tilth), there are a number of materials and products that loosen "dead," compacted soil. They may also help eliminate a hardpan. By loosening soil and allowing air to enter, beneficial soil organisms can flourish, and healthier, better quality crops are the result. Some of the products do add fertilizer nutrients themselves, but others contain few nutrients and serve only to improve tilth, so are usually called *soil conditioners*.

1. First we have various types of natural mined and ground-up rock deposits. They help particles of "tight" soils to cling together to form a desirable "crumb structure" ("good tilth"), plus they may contain significant amounts of secondary and trace elements.

A. *Liming materials.* Ground high-calcium limestone (calcium carbonate) and other high-calcium liming materials (marl, water treatment plant lime, papermill waste, slags) can greatly improve structure in many types of soils, especially eastern acidic, clayey soils. In the western alkaline soils, gypsum (calcium sulfate) may improve soil structure. Certain mined products from the western U.S. that contain large amounts of gypsum shale have been found to improve soil structure and supply trace elements. Dolomitic lime (calcium magnesium carbonate) is an extremely hard stone and slow to become available to plants. Also, its high magnesium content may provide excessive amounts of this element if it is not needed.

B. *Humates.* Humates are mined near commercial coal deposits in the western U.S. They are called oxidized lignite and are high in salts of humic and fulvic acids. These acids are also found in humus and compost. They have been found to be plant growth stimulants and can chelate trace elements, making them more available to plants. At moderate rates of application (500–750 lbs./acre broadcast), humates can improve soil structure and drainage, rooting, growth, and crop quality, including shelf-life of fruits and vegetables. However, too much humic acid will inhibit plant growth, so it is best to keep application rates low (less than 100 lbs./acre broadcast or 15–30 lbs./acre in the row). Humates work best when mixed with fertilizers. They will have little noticeable effect in soil with high organic matter.

C. *Rock phosphates.* There are two common types of rock phosphate products. Ordinary ground rock phosphate, sometimes called hard rock phosphate, is mined mostly for acid treatment to make superphosphate.

Some is sold for use by organic farmers and gardeners. When the rock phosphate is ground and washed, the washings are allowed to settle in settling ponds in the form of a mixture of small phosphate particles and clay. This material is also sold, sometimes called soft rock phosphate or colloidal phosphate.

The hard rock phosphate takes some time for its nutrients to become available to plants. The soft rock phosphate, being colloidal, is a quickly available source of phosphorus, a small percentage at a time. But it also helps condition soil and prevents leaching of lime. It works well along with high-calcium lime, in a ratio of about one-half as much phosphate per acre as lime. A test in Georgia showed that rock phosphate can give forage yields equal to superphosphate, over a five-year period.

2. Next we have *biological products*, either living organisms or substances produced by them. Some of them not only improve soil structure, but also release plant growth stimulants.

A. *Microorganisms.* It is well known that many soil bacteria and algae produce sticky carbohydrates (polysaccharides) that glue soil particles together, giving good soil structure. A healthy, "living" soil should have plenty of microorganisms, but "dead" soil may have to be inoculated or stimulated. Some commercial products contain mixtures of bacteria species or bacteria plus fungi and yeasts. Some contain algae. Besides helping condition soil, certain microorganisms also fix nitrogen from the air, reducing fertilizer needs. Some commercial products contain microbial activators, mixtures of nutrients, enzymes, and other materials, intended to stimulate the growth of existing soil life.

Getting new species of microorganisms established in soil is not always successful, depending on soil conditions and substances toxic to the organisms. Also, sometimes species already present will out-compete or kill-off the introduced species. Such products can give good results or none at all. It has been found that just spraying about a quart of molasses (diluted in water) per acre on soil or applying about one-half to one pound of granular sugar per acre improves soil structure and stimulates soil life by serving as a food source for organisms. No more than 3 gallons per acre per year of molasses should be used. Adding any type of organic matter, if it is kept in the upper aerated layers of the soil, will stimulate microorganisms by serving as a food source.

B. *Organic matter.* Organic matter, either fresh or decomposed into compost or humus, does an excellent job of improving soil structure. Large amounts of fresh organic matter should not be incorporated into the soil because toxic by-products of microbial decomposition could be formed. Such materials as crop residues, animal manures, green manure, cannery wastes, sewage sludge, etc., are good sources. A mixture of animal and plant sources gives better results, and adding ground rock fertilizers is even better. Fresh organic matter should readily decompose to humus in the soil. The other alternative is to apply composts, which do not have the danger of toxicity. Besides improving soil structure, organic matter and compost increase soil life and supply many plant nutrients. Some companies specialize in compost technology and sell compost, compost starters (bacterial inoculants), and compost-turning machinery. See Chapter 13.

Kelp or seaweed is one type of organic matter readily available near the oceans. If incorporated into the soil, it will not only condition soil, but also supply trace elements and growth-stimulating hormones (when used at rates of 250–500 lbs./acre, vegetable yields generally increased from about 10 to over 100% in some tests). Some companies sell seaweed meal or powder for farmers far from an ocean. Research at the University of Maryland found that seaweed meal on lawns greatly reduced nematode populations.

C. *Enzymes.* Soil sprays containing microbial-produced enzymes have been found to improve soil structure, eliminate hardpans, correct extreme pH, increase the soil's nutrient-holding capacity (CEC, cation exchange capacity), and detoxify soil, as well as increase yields and quality.

3. Finally, there are the *wetting agents,* sometimes called *surfactants.* A surfactant is a chemical that lowers the surface tension of water. A soap or detergent is a familiar example. There are a number of surfactant spray products for use on soil, which can increase the rate of soaking-in of water, overcome a hardpan, and drain wet spots. Research has shown that they work well on hydrophobic (water-repelling) soils, but not on hydrophilic (wettable) soils. Generally, water-repellent soils are fine-textured. You can test this by placing a drop of water on the surface; if it beads up and soaks in slowly, a wetting agent may help your soil. Wetting agents should be used in moderation as a "band-aid" until you get more humus and microorganisms into the soil because they can adversely change soil structure, reduce crumb structure and cause microorganisms to lose their protected homes inside soil clumps.

Chapter 18

QUALITY & HEALTH

W hich would *you* rather eat, a nice plump, red, juicy, sweet apple or a shriveled, green, bitter, half-rotten thing called an apple by your grocer? The answer is obvious; you want quality food for yourself. So why do most farmers feed miserably poor quality, vitamin and mineral deficient, and even toxic feed to their animals? And then they have to add supplements to fortify it and molasses to get the critters to even eat it!

The general trend of western civilization seems to be away from quality and toward "just enough to get by." This is clearly indicated by the flood of cheap plastic throwaway products on the market. The ultra-competitive climate fostered by modern capitalism causes corporations to use the cheapest possible materials and labor, in order to maximize profits and keep their stockholders happy. Even in agriculture, the emphasis for so many years has been on quantity—bushels and pounds—and not quality.

Dr. William Albrecht wrote that in 1911, the protein content of corn was 10.3% (this was before hybrids). In 1950, five hybrid varieties were tested; the protein content ranged from 7.9 to 8.8%. In 1956, 50 hybrid varieties averaged 5.15%. The same trend has occurred in other crops; in 1940, Kansas wheat averaged nearly 19% protein; it dropped to 14% in 1951 and 10.5% in 1969. These are shocking losses in quality. At the same time, yield per acre has steadily risen. Some average yield figures from the *Statistical Abstract of the U.S.* and the USDA *Agricultural Statistics* are:

Corn, bushels/acre		Wheat, bushels/acre	
1931–35	22.8	1931–35	13.1
1941–45	32.8	1941–45	17.5
1951–55	39.1	1951–55	17.9
1957	47.1	1957	21.7
1960	54.7	1960	26.1
1966	73.1	1966	26.3
1969	85.9	1969	30.6
1973	91.3	1973	31.6
1979	109.7	1979	34.2
1985	118.0	1985	37.5
1990	118.5	1990	39.5
2000	136.9	2000	42.0

Such are the marvels of modern agriculture: more bushels per acre, lower nutritional value, sick animals, higher veterinary bills!

What do we mean by quality? How is quality measured? Is high quality really important? Why take the trouble to strive for quality?

Quality—what is it? The dictionary defines quality as degree of excellence; superiority; capability of producing specific effects. That can certainly be applied to agriculture, but it depends on what criteria are chosen to measure quality. Some are very misleading. For example, animal nutritionists and feeds experts usually measure crude protein (CP). This may sound like an overall measure of protein content, but in practice, the lab technician only measures the % of nitrogen in the ashes of the sample and multiplies by 6.25 to give "crude protein." This method is based on the *assumption* that the protein content (which is harder to measure) is 6.25 times greater than the nitrogen content. This may or may not be true (Research Report 29, Part 1, Univ. of Wisconsin College of Agriculture Experiment Station, Oct. 1967, p. 2). If the metabolism of the plant was upset, protein synthesis may have been slowed or blocked, and substances preliminary to protein formation may have built up in the plant's tissues: nitrate, ammonium, amides, and amino acids. These would give a higher nitrogen reading, but they are not protein. And protein content isn't the whole story. The proper balance of amino acids in the proteins is necessary for good health in animals.

We have covered some methods of measuring nutritional value in animal feeds in Chapter 5. There are many methods, some easy, some hard, some better, some misleading. Yield in bushels or pounds per acre is of course not a reliable measure at all. That is why we are growing bumper crops of "foodless food." A somewhat better measure is test weight. For decades the standard weight of a bushel of corn was 56 pounds. Then so much poor quality corn was being grown that the weight of number 2 corn was lowered—and lowered—to 53 pounds—and then 52 pounds. No, bushels haven't gotten smaller, corn is getting lighter. Maybe we're growing cornflakes! Yet *some* farmers are growing 60 or 61 pound corn. They have their soil in balance.

The best measure of food quality is what it does for the animal (or person) that eats it. Does it cause good weight gain, milk or egg or meat production, normal reproduction, vigor and disease resistance? Decades ago, the standard amount of feed needed daily by an average dairy cow was 32 pounds. Today, 50 or more pounds are considered necessary. Maybe more is needed for greater milk production, but why are protein and mineral supplements "necessary" now? Are we growing more bulk "belly filler"—and not nutritious feed? Farmers with in-balance soil report that their cattle eat only one-half or two-thirds as much feed, but give more milk with higher test figures, and the incidence of mastitis and other ailments drops or disappears completely. For example, a northeastern Wisconsin dairyman, after beginning to feed his herd feed grown on balanced soil, was pleased to find his average butterfat figure rise from 540 pounds in May of 1982 to 555 pounds in July, and his rolling herd average go from 15,260 pounds in May to 15,499 pounds in July.

You can measure it. Technology-orientated consultants and farmers can use an infrared digital camera to detect subtle color changes in sick plants, and certain orange-tinted glasses can do the same. Researchers are developing mobile robots that can scout fields for nitrogen deficiency, water stress, weeds, diseases and pests.

A quick, easy way to get an indication of crop quality while the plants are still growing is to use an instrument called a *refractometer*. A refractometer is a relatively inexpensive hand-held device which measures the sugar content (technically, the soluble solids—primarily sugar) of a couple drops of any liquid. Refractometers are commonly used in industry as a quality check: by sugar beet buyers to determine grade and price paid to growers, by wine makers to check grape quality, and by alcohol distillers to test grain. They are used by some doctors and veterinarians to test urine. You simply squeeze a few drops of juice from the stem or leaves of your crop plant onto a glass prism in the refractometer and then look through the eyepiece. The sugar content is read on a numbered scale in units called Brix (same as percent). By comparing with standard levels (see box) and with past readings for your crops, you can see how well your crops are functioning—making sugar. You should be careful to test juice from the same part of a plant each time, say the corn stem at the third leaf, or at the cob level, since sugar content does vary along the length of the stalk. Also, readings will be higher on sunny days. An unhealthy (stressed) plant may have an accumulation of sugars in its leaves or stems, giving a deceptively high refractometer reading. You can usually see signs that the plants are abnormal, such as stunted growth or purple color. A healthy plant should not have high sugars in the early morning in its upper leaves or stem.

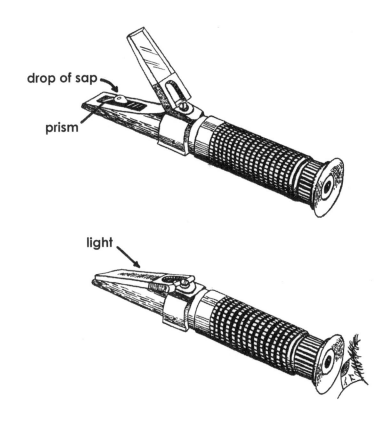

drop of sap

prism

light

Using a refractometer is easy. A drop or two of sap from the plant is squeezed onto the glass prism. Then the "lid" is closed; the instrument is held in the light, and the sugar content is read on a scale visible through the eyepiece.

Although the refractometer only measures sugar content, it is still a good (and easy) way to measure quality and feed value, since tests have shown that protein and mineral content are directly correlated to Brix readings. Generally, the highest refractometer reading is the best time to harvest a crop. You can even use the refractometer to test the quality of fruit and vegetables you buy at the store (although the grocer may take a dim view of it).

Still another good indication of quality for many crops is their storage and keeping ability. People who store fruit and vegetables certainly know that in some years, produce keeps better than others or that crops from certain fields store better. If grain isn't dry enough, it will mold in storage. Generally today, corn quality is so low that farmers haul their corn to the

Approximate Brix (sugar) Readings for Some Plants

	poor	average	good	excellent
alfalfa	4	8	16	22
carrots	4	6	12	18
grains	6	10	14	18
corn stalks	4	8	14	20
corn, young plant	6	10	18	24
field peas	4	6	10	12
lettuce	4	6	8	10
sorghum	6	10	22	30
soybean	4	6	10	12
sweet corn	6	10	18	24
tomato	4	6	8	12
watermelon	8	12	14	16

local mill or elevator and pay to have it dried by using expensive fossil fuel. How much easier it is to simply "grow" your grain dry; that is, to have healthy plants which finish the normal physiological development of the grain, so that the water content of the maturing grain is naturally reduced at the same time that the kernel is filling up with food and minerals. Such completely matured grain will store mold-free at a higher moisture content than incompletely matured grain (what moisture there is in mature grain is "bound moisture," locked into the stored food molecules; it will not cause mold problems). In order for this to occur, the soil and plant have to be in top condition at the end of the growing season, and the "plumbing" of the plant must not be plugged up or degenerated (see Chapter 4). Believe it or not, corn doesn't have to dent; even "dent" varieties will not dent as much as usual when they are of high quality and full of minerals and protein. Moisture content and storage quality are correlated with the sugar content, easily measured with a refractometer.

But besides all these ways for man to measure quality of crops, probably the best overall "instrument" to let you know the quality of your crops is your animals. As Dr. William Albrecht has put it, "Cows are capable chemists." Animals have much better senses of smell and taste than we do, and they have the instinct to "know" what nutrients they need. Every farmer who feeds livestock or poultry has probably noticed that animals prefer some feeds over others, or hay from one field.

A study in Australia found that sheep ate more clover hay from test plots with higher phosphorus contents (*Australian Journal of Agricultural Research*, 1971, p. 941–950), and another study found that sheep preferred tall fescue hay fertilized with phosphate and moderate levels (50 pounds per acre) of nitrogen (*Journal of Animal Science*, 1965, p. 615–625).

Dr. Albrecht points out that cattle avoid eating the taller and greener grass growing around manure "pies" because in that small area, the plants' fertility is out-of-balance, with *too much* nitrogen being provided by the manure. The lush grass which looks better to us is actually of lower nutritional quality to the animal. Dr. C.D. Story of the Colorado State University Agricultural Experiment Station, found that "animals fed nitrogen-fertilized hay do not make as high gains as those fed unfertilized hay" because "fertilizers that provide quickly soluble nitrogen will lower the capacity of these crops, hybrid crops in particular, to produce seeds with high-protein content" (*Organic Gardening and Farming*, May 1960). Cattle will often graze next to a concrete or gravel road because the additional lime from the road improves forage quality. They even prefer to eat "weeds" growing on fertile soil over grass on poor soil (W.A. Albrecht, *Soil Fertility and Animal Health*, 1958, p. 113).

Are hybrids really better? One of the supposedly great boons to agriculture from applied science is the "improvement" of plants and animals from crossing different stocks or lines to obtain hybrids. These new hybrid varieties often have vastly different characteristics than their parents—greater yields, disease resistance, shorter maturity time, etc. Plant and animal breeding have brought us such wonders as tomatoes that are designed to be of uniform size and to be firm enough to be able to be picked and packaged by machines—but they taste like cardboard!

Hybrids result from applying the laws of genetics to obtain a desired combination of characteristics in an offspring. The physical and functional characteristics of all organisms are controlled by their genes. Different

Aflatoxin

Aflatoxins are several highly toxic substances produced by two common molds, *Aspergillus flavus* and *Aspergillus parasiticus*, when they grow on certain foods and feeds. The name comes from *Aspergillus flavus-toxin*. One type of aflatoxin, B_1, is the most powerful cancer-causing agent known. Even at the lowest detectable amount, 0.1 parts per billion, scientists are concerned about its hazards in human and animal foods. Other less dangerous toxins are produced by certain other fungi.

Aflatoxins are produced when the two *Aspergillus* species grow under warm, moist conditions on various crops: peanuts, corn, wheat, oats, barley, rice, sorghum, tree nuts (pecans, Brazil nuts, pistachio nuts), coffee, black pepper, and feed meals (cottonseed meal, soybean meal, linseed meal, and fish meal). These foods can be contaminated while in storage or crops can become moldy in the field. Aflatoxins are mainly found in tropical and subtropical areas, and in the U.S. are most common in the Southeast. Abnormal or unhealthy crops are most often attacked: cracked kernels of stored high moisture grain, or crops damaged by hail or early freeze, or suffering stress from drought and insect damage. In 1977, there was a serious outbreak of aflatoxin fungi in the Southeast on corn stressed by drought and insects; as much as two-thirds of the crop in some states was contaminated.

The danger of aflatoxin to animals and humans was not recognized until 1960. In larger doses animal poisoning and other side effects result. Symptoms include impaired liver functions, reduced blood clotting, fragile capillaries and hemorrhaging, kidney damage, impaired immunity to diseases ("lazy" white blood cells which do not destroy bacteria as they should), interference with vitamins A and D and with calcium metabolism, anemia, reduced weight gain and production of milk or eggs, susceptibility to diseases (pneumonia, mastitis, foot rot, emphysema), and poor reproduction (low conception rate, false pregnancies, abortions, and still births). A long low-level exposure results in liver cancer in more susceptible animals.

In 1960, thousands of young turkeys, pheasants, and ducks in England died; the cause was traced to a shipment of Brazilian peanut meal used as a protein feed supplement. In 1974 in India, 400 people were poisoned by eating contaminated corn; 106 died. Unusually high liver cancer rates in Africa, India, and the Philippines are thought to be connected with aflatoxins ("Aflatoxin and Other Mycotoxins: An Agricultural Perspective," Report No. 80, Council for Agricultural Science & Technology, 1979).

Aflatoxins cannot be detected by ordinary means. Not all moldy grain or feed contains them. A quick screening check is often run by examination under ultraviolet light; aflatoxin-contaminated materials will glow greenish-yellow, but a further chemical analysis is required for a positive identification.

When animals consume aflatoxin-contaminated feed, most of the toxin is excreted, but some will appear in milk or eggs, and some will be stored in the liver and muscle. It can also find its way into cheese and other dairy products. Sweet corn (fresh, frozen, or canned) is not susceptible to contamination, nor is corn starch or refined corn or peanut oils, but other corn and peanut products are suspect (cornmeal, grits, etc.).

Aflatoxins are difficult to remove once produced. High heat treatments, dissolving out with alcohol or acetone, treatment with strong acids or alkalies, and inactivation with ammonia, formaldehyde, hydrogen peroxide, or other chemicals are some methods that have worked. Chemicals that kill the fungi will prevent aflatoxins from being produced. Research is being done to find crop varieties that are resistant to fungal attack. There are natural substances in soybeans and some peanuts that reduce attack. Grain and meal should be stored under conditions that do not favor fungal growth. Contaminated feed is sometimes blended with good grain to reduce the aflatoxin to a "safe" level and then fed to "resistant" animals. Since only damaged or stressed crops are affected, doesn't it make sense to build up soil fertility and the health of the plant?

individuals of a species carry different traits in their set of genes (there are thousands of individual genes). As we all know from human heredity, the children may or may not inherit dad's nose or grandma's curly hair. Geneticists identify "desirable" traits in various varieties of wild and domesticated plants and animals, then breed a pure inbred line having a certain trait. They then cross parents from two inbred lines to obtain a hybrid which they hope will have a certain combination of traits. Sometimes two hybrid lines (single crosses) will be crossed to give a double cross. It is a difficult and time-consuming process, although a new method, called marker-assisted breeding, promises to speed things up by using DNA fingerprinting technology to mark desirable genes. Now scientists are using sophisticated techniques to transfer and control single genes from other species to give desired characteristics, called genetic engineering or gene modification (see Chapter 1 and box, below).

GMOs—Friend or Foe?

Genetically modified organisms (GMOs) were researched, developed, patented and released commercially in an exuberant flood of promising propaganda by a few big agribusiness companies, led by Monsanto. GM (genetically modified) crops are necessary to "wipe out hunger" or "feed the world," we were told. These new crops will protect the environment by reducing the amount of toxic herbicides and insecticides used. Therefore, with lower input costs, GM crops will make huge profits for beleaguered farmers. And then the use of genetically engineered plants and animals to produce a host of pharmaceutical products will save countless lives. So we were told.

GMOs aren't "mutant monsters"; they are ordinary plants or animals, just with a little DNA from somewhere else. They won't, or shouldn't, escape or overtake the world, we were told.

O.K., so GMOs were given a terrific send-off. Then what happened? Have GMOs lived up to the promises? First, what about the supposed safety and lack of spread of GM genes to other organisms? Disturbing accounts are leaking out that some of the company-sponsored research to prove the safety of their GM products was either poorly performed or downright rigged.

Tests run by unbiased scientists have found that when mice or rats are fed GM tomatoes, corn, soybeans or potatoes, they develop pre-cancerous cell growth, crippled immune systems, damaged livers, and abnormal development of brain, blood cells, liver and testes (in many experiments, the rodents didn't even want to eat the GM food). Tests with pigs, cattle and chickens also found similar health problems, as well as sterility and premature death.

Galloping genes. Once unleashed into the environment, the foreign genes in GMOs can easily spread to other organisms. One way is by GM pollen blowing around and pollinating non-GM crops, as has happened for canola and corn. Also, it has been found that when GM corn pollen carrying the gene for *Bt* lands on milkweed plants or monarch butterfly caterpillars, the insects can be killed, since *Bt* kills any caterpillar, not just corn borers as intended. Foreign genes that were inserted into the GMO by bacteria or viruses (which invade the cells of the GMO), can just as easily invade the cells of non-GM organisms, called horizontal gene transfer. When human subjects were fed GM food, the foreign genes were found in the normal bacteria in their intestine.

Still another safety problem is that when foreign genes are inserted into the cells of an organism, its gene contents are more or less scrambled with unpredictable results; in normal cells, various genes interact with and modify one another, and their closeness or lack thereof makes a lot of difference. Also, so-called promotor genes, which switch other genes on or off, are included in the gene transfer, and they can produce uncontrolled cell growth or cancer in the receiving species. Any extraneous genes from the carrier bacteria or virus can also have unknown effects, mutating at random and possibly causing toxicity or dangerous new allergies. Severe human allergies can also be caused by allergy-producing genes from the donor species (as happened when a gene from Brazil nuts was put into soybeans) or from allergy-protein amplification in the recipient species. In 2000, the GM corn variety Starlink™, used for animal feed, had to be discontinued because it accidentally ended up in human food and caused allergies.

Superpests. Whether the use of GMOs to produce useful drugs, etc., will prove beneficial is still uncertain. It could be genetic engineering's only saving grace. But there are problems. When scientists transfer genes from one species to another, they include a marker gene that produces resistance to certain antibiotics, so they can test whether the gene transfer was successful. Any person or animal that eats them in their food risks the possibility that new antibiotic-resistant microbes can be produced. Bacteria and viruses readily exchange genes among themselves, and the next pandemic could result. Or, as a result of planting millions of acres with herbicide-resistant and insect pest-resistant crops, it is very likely that "super-weeds" and "super-pests" will develop that are resistant to herbicides and the *Bt* toxin. A few examples of this have already been found.

More, not less. The promise of reduced herbicide and pesticide use haven't held up. Studies published in 2002 and 2003 on the results of growing GM crops have found either very slight herbicide/pesticide reductions, or more often considerable increases. For example, herbicide-resistant soybeans used over 50% more herbicide (active ingredient) per acre than non-GM soybeans. In the years since GM crops have been grown in the U.S. (until 2003), GM crops have increased herbicide/pesticide use over 50 million pounds. More toxic chemicals in the environment mean more death and ecological damage to wildlife and humans.

Profit or loss? In an analysis of the economics of growing GM corn, between 1996 and 2001, U.S. farmers paid at least $659 million for the expensive seed, but received only $567 million for their product, a loss of $92 million.

And as we have mentioned, GM crops overall do not produce greatly higher yields than ordinary crops, and their nutritional content is often substandard; thus GM crops will not save the world from hunger, but they make the companies that sell them very fat!

Who wants them? Still another problem of genetically modified crops and food is that enough right-thinking people are aware of the draw-backs and dangers posed by GMOs so that a large majority of consumers and nations don't want any part of them!

Opinion polls of Americans found that up to 94% of respondents said they wanted food labels to tell about any GMO ingredients, and about 60% said they would avoid such labeled products. Japan, Australia, New Zealand and Canada have banned import of GM foods, and the European Union has a moratorium on them, although they are starting to worm their way into some markets. Europe has required labeling of GM foods for many years. This refusal of foreigners to accept GM foods has put quite a bind on the growers of the crops and the companies that promote them.

Lab rats. Around 75–80% of processed food products in the U.S. contain GM ingredients. We eat them every day! Yet the U.S. Food and Drug Administration (FDA) does not require product labeling or safety testing of GMO-containing foods. We are "lab rats" in a great uncontrolled food experiment. Unlike most other developed countries, there is little government oversight of the biotech industry, and they predict that nearly 100% of U.S. food and fiber products will be genetically engineered by 2020 or so. According to Jeffrey Smith, author of *Seeds of Deception*, a biotech insider has said that the industry is hoping that GMOs will eventually contaminate the whole world, and then the technology's critics will finally give up. One method of achieving this is shipping GM seeds in food aid to starving countries. Unless the grain is milled, poor farmers are likely to plant some, and Pandora's box will be open. Some futurists and scientists want to use genetic engineering to improve the human race, both mentally and physically, to eliminate diseases and aging. This is the latest incarnation of the "transhumanism" movement from the early 20th century. The possibilities are chilling.

The bottom line. Biotechnology and genetic engineering are the latest revolution in agriculture and food production. It was an unexpected outgrowth of rapid scientific development in the 20th century. Creative scientists are usually willing to try just about any feat, just because they can. Giant corporations with the capital to fund high-tech research are eager to develop and market new products . . . with the goal of a fat bottom line. They are often aided by government agencies and politicians who value profits ahead of the public good.

Biotech companies use tactics of intimidation against those who cross them, including farmers who are found growing crops containing the company's patented genes without permission. The biotech industry seems unstoppable; it is a monstrous threat to humanity and planet earth. Let's hope it can be stopped. And that's the bottom line.

Is all this "tinkering" with nature really good? Some amazing things can be done, and at first glance, it seems great to have corn that is resistant to southern corn leaf blight and cows that give more milk with higher butterfat. One obvious disadvantage of hybrids is that they do not "breed true"; that is, planting seed from hybrid plants does not produce all hybrids in the next generation because some of the offspring revert back to their ancestral traits. So the farmer is "locked in" to buying hybrid seed every year. Crossbred cattle may have good characteristics, but in the next generation, many undesirable offspring are born (*Western Livestock Journal*, Oct. 1969, p. 25–26).

Another problem with hybrids is that they are too limited in their characteristics—they are "fine tuned." They are all alike for certain traits (within a given variety), and there is no variation. While certain "desirable" traits are present (are they always really desirable?), others are therefore missing, which would give the plants or animals more adaptability to varying conditions—weather, nutrition, pests. Natural populations of plant and animal species have a variety of gene traits in the different individuals; while a few may be unfit, most will be well adapted for their environment. An excellent example of the vulnerability of uniform hybrids is the wildfire spread of southern corn leaf blight in 1970. Because of these things, scientists are becoming concerned about the "genetic vulnerability" or "narrowing genetic bases" of our crops through hybridization, and the possible loss of the original wild species of plants (from extinction by man's activities) from which the domesticated varieties were bred (see Chapter 1). "Gene banks" of stored seeds from many native varieties of crop plants are now being maintained in order to preserve genetic traits that may prove useful in the future.

What are the "desirable" traits that agricultural geneticists seek for in hybrids? Too often they emphasize high yields—quantity, not quality. While there are certain strengths, there are also weaknesses. If you press a hybrid corn breeder, he will admit that the various varieties are "fine tuned" to perform well on different soils—soils with one or another deficiency or fertility imbalance, such as high magnesium or high potassium, and some will even say that they do not have a variety that will do well on "good" soil. Tests by Ernest M. Halbleib of McNabb, Illinois, have found that hybrid corn did not absorb 7 to 9 trace minerals. None of the varieties Halbleib tested took up cobalt, a mineral needed in vitamin B_{12}, which animals need for good health. A cobalt deficiency has been linked to the development of brucellosis and undulant fever. According to the University of Nebraska Experiment Station Bulletin 144 (1946), open pollinated corn makes chicks gain faster and have more protein in their meat than hybrids, and it makes far superior quality silage.

The "green revolution" of the late 1960s seemed like an agricultural savior at first glance. The new hybrid varieties of corn, wheat, and rice, produced greatly increased yields—but at what cost? They could not grow successfully unless given large doses of fertilizers and protected by pesticides, which underdeveloped countries cannot afford (see Chapter 1).

Theoretically, not all hybrids are bad. It just depends on what particular combination of traits (genes) wind up in the hybrid variety. Do they make the plant or animal more vigorous and give more nutritious food, or do they produce an out-of-balance plant or animal that gives "foodless food"?

A growing number of farmers are finding that open pollinated crops can produce yields nearly equal to hybrids, but of much higher quality. Adolph Steinbronn of Fairbanks, Iowa, had his open pollinated corn tested. Compared to average Midwestern test figures, Steinbronn's corn had 75% more crude protein, 875% more copper, 345% more iron, and 205% more manganese (*Eco-Farm—An Acres U.S.A. Primer*, 2003, p. 38). It all traces back to the soil.

Quality—the role of fertile soil. We have seen in Chapter 16 that soil fertility—good or poor—can affect the resistance of plants to diseases and pests. This happens through internal functions and processes of the plant; that is, its metabolic functions are different. Is it any wonder then that soil fertility can also influence the food quality of crops—their protein, vitamin, and mineral content?

Even though some "experts" have claimed that soil fertility does not affect the vitamin or nutrient content of crops, numerous studies have proved that it does. Tests run on vegetables from different areas have found a four-fold phosphorus variation and a 32-fold iron variation in spinach. Tomatoes have been found to vary over two times (200%) in phosphorus content, nearly three times in potassium, five times in calcium, over six

Poor and good root development in corn from the same field, but under different fertilizer programs.

times in boron and cobalt, 12 times in magnesium, 53 times in copper, 68 times in manganese, and 1938 times in iron (J.I. Rodale, *The Complete Book of Composting*, 1972, p. 595–596). Dr. William D. McElroy of Johns Hopkins University found that unbalanced soil trace element nutrition can interfere with a plant's enzyme system and make it inferior in quality long before it appears stunted or shows leaf discoloration. He found that iron and copper deficiency reduced enzyme activity in tomatoes by over 80% (*The Complete Book of Composting*, p. 612). Dr. Charles Northern has found that adding a small amount of manganese increased the vitamin C content of plants three times, while boron increased vitamin A in apples 100% (*The Complete Book of Composting*, p. 614). Experiments growing peppermint in complete and deficient nutrient solutions showed that the amino acids produced by plants are unbalanced if the plant's nutrition is deficient (G.R. Noggle & G.J. Fritz, *Introductory Plant Physiology*, 1976, p. 251; see box).

High levels of potassium or low levels of phosphorus increase the nitrate content of plants, while balanced fertilizer levels produce highest quality silage (Research Report 29, Part 1, Univ. of Wisconsin College of Agriculture Experiment Station, Oct., 1967, p. 2, 7).

Besides the beneficial effects on the physical characteristics of the soil (tilth, aeration), organic matter (humus and the microorganisms living

PERCENT OF ALCOHOL-SOLUBLE AMINO ACIDS AND AMIDES IN LEAVES OF PEPPERMINT GROWN IN VARYING NUTRIENT SOLUTIONS

amino acid or amide	complete nutrients	potassium deficient	calcium deficient	phosphorus deficient	sulfur deficient
glutamine	12.7	50.1	6.0	18.7	9.2
leucine	1.8	6.7	1.8	1.2	—
valine	3.1	4.3	2.4	2.5	1.2
alanine	6.0	4.7	4.7	3.2	1.3
serine	3.6	3.6	2.1	3.3	1.2
S-amino butyric acid	6.7	2.8	5.6	3.6	5.6
glutamic acid	38.8	3.5	25.4	30.7	3.3
aspartic acid	27.3	6.6	10.3	26.0	3.3
asparagine	—	15.6	29.2	10.8	16.4
tyrosine	—	2.1	—	—	—
proline	—	—	9.7	—	—
arginine	—	—	—	—	58.5
methionine	—	—	2.8	—	—

in it) has a great deal to do with producing high quality crops. Growth-promoting substances are produced in the soil, trace elements are supplied to plants in more available forms, and crops contain more vitamins than those grown without organic matter.

There is no substitute for healthy, living soil. Without it, truly nutritious, high quality crops cannot be grown, and healthy, high-producing animals cannot be raised.

Calcium-Potassium

• From W.A. Albrecht, "Soil Fertility as a Pattern of Possible Deficiencies," *Journal of the American Academy of Applied Nutrition*, Vol. 1, 1947, p. 17–28:

"The calcium-potassium ratio . . . has given us a pattern of the protein possibilities in the crop. If nature, under less rainfall, has left much calcium in the soil, we have a proteinaceous crop. If the soil is under higher rainfall to give a small amount of calcium in relation to potash, then we have a carbonaceous crop. The validity of this belief, namely that a liberal supply of calcium in the soil in relation to potassium represents production of crops rich in proteins and minerals, while the reverse relation give crops high in carbohydrates—thereby low in proteins and minerals—was tested. Soybeans were grown with increasing amounts of potassium available in the soil and associated with constant amounts of calcium. Three ratios of calcium in potassium were used while all other nutrients were liberally supplied. Increasing the potassium increased the forage yield to a maximum of 25%. This fact would draw ready applause for an experimenting agronomist. Such work can win funds in support of it as research. But the buffalo of the western plains didn't evaluate herbage in terms of bulk. Our livestock does not use that criterion either. Hence, while increase of bulk may appear laudable, fixing our attention on bulk in relation to soil fertility has been leading us to grow more crops with serious deficiencies as feeds.

"Chemical analyses were made of the forage. The nitrogen content of the smallest of the three crops was 2.8%; of the intermediate crop

2.5%; and of the largest crop 2.19%. While we increased the bulk 25%, we reduced the concentration of nitrogen, and therefore the protein, by more than that figure. So that the greater amount of total protein was not in the largest but in the smallest crop.

"The phosphorus concentration, by analysis, was .25% in the small crop; .18% in the intermediate; and .14% in the largest crop. Assuming that the cow could digest it completely, she would be compelled to eat approximately twice as much of the larger crop to get the same amount of phosphorus. In the case of the calcium, this was approximately .75% of the dry weight in the smallest crop and only .27% or about one-third as much in the largest crop. Can any cow increase her consuming capacity by three times? We can't expect her to become a hay baler.

"We need to be concerned not only with the bulk of the crop, but also with the synthetic operations of the plant in using the fertility elements from the soil to convert the carbonaceousness over into proteinaceousness. Those processes make "grow" foods instead of merely "go" foods. They must be more generally appreciated if we are not to invite nutritional deficiencies more commonly."

• From Dr. Frank A. Gilbert, in his book, *Mineral Nutrition of Plants and Animals* (1948), p. 43, 45:
"The present opinion seems to be that a high potassium content of the soil when not balanced by other essential elements, especially calcium, tends to produce a plant especially high in carbohydrates. A normal amount when calcium and phosphorus are deficient will have the same effect and will produce plants of a very low biological value with a lesser change in caloric value. On the other hand, a more nutritious proteinaceous and less carbonaceous plant grows when the potassium-calcium ratio is weighted in favor of the latter element."

Soil health. We are used to using the term *health* for ourselves, for our animals and for plants, but since the early 1990s agronomists and soil scientists have developed the concept of *soil health* or *soil quality*. The idea of soil quality was originated to encompass the concepts of agricultural sustainability, soil's capacity or fitness to produce plants (and animals), and its ability to resist and recover from degradation. There were previous methods of classifying soils according to their physical and chemical (pH and nutrients) characteristics, but the dynamic aspects of biological life and its influence on organic matter recycling and plant functioning were mostly ignored.

No doubt for hundreds of years, farmers and gardeners have had a general, perhaps nebulous, idea of what is "good" soil and what isn't, probably based mostly on how well it produces "good" crops. Years of observations of how easily their fields can be plowed, how quickly rainfall drains, whether a crust forms in dry weather, and so on, contributed to a sense of good and poor soils. In general, a "healthy" soil has a good balance of air and water, a good balance of plant nutrients, a fairly high level of organic matter (humus), and a high level of beneficial soil organisms (microbes, earthworms, etc.).

But scientists love to measure things; they hate generalities. Thus various soil quality assessments, scorecards and test kits have been developed to help researchers, consultants and farmers determine the status of any field or plot. These methods combine tests and observations from three areas: *physical* (such as density, stability of soil aggregates or crumbs, water/air content, and rate of water infiltration), *chemical* (pH, electrical conductivity, and amount of certain nutrients), and *biological* (earthworm count, organic matter level, and soil respiration, which measures microbial activity). All of the measurements are put through a computer and out comes an overall soil rating, or soil quality index, a number score. So you can say that your valley field that has an index of 7.2 is a lot better than that hilltop field which only rates 3.8.

But while an overall numerical score has some comparative value, does it really tell you much that is useful? It is more important to know exactly what component or aspect of your soil is good or bad. That way you can know what needs to be corrected. The best thing about the interest by scientists in overall soil quality is just that—they are starting to think holistically and to realize that soil fertility and health doesn't just consist of a few physical and chemical aspects such as pH and N-P-K, but that the complex functions of soil organisms and humus are also vitally important.

Why bother? O.K., you say, so high quality crops are better. But the market pays for bushels and pounds. What's the use of growing high quality crops if I can't sell them for a good price?

Of course the value of growing high quality crops is obvious if you are going to feed them to your own livestock or eat them yourself. The problem comes when you raise crops to sell. Unfortunately, the general market is mainly only interested in bulk, in quantity, not quality. So where can a quality-conscious farmer find markets?

Markets. First of all, some markets do pay a premium price for high quality, such as for Grade A milk or high test weight grain, but you should be able to get better returns by seeking non-conventional markets. There are a surprising number and variety, including:

1. Organic food outlets and CO-OPs. If you farm entirely without synthetic chemicals, you may be able to become certified as an organic grower. Each state or region has its organic marketing network.

2. Supermarkets, health food stores, bakeries, restaurants, etc. Many food wholesalers and retailers are interested in high quality products. Ask in your local area. Many up-scale restaurants seek out and work directly with local organic or sustainable growers. Breakfast cereal manufacturers want quality grain.

3. Foreign markets. Most foreign grain importers definitely want high quality grain and will pay premium prices. You will need to make contacts to deal directly with the foreign buyers and bypass domestic markets.

4. Bird or animal food. Poultry raisers, egg farmers, zoos and bird seed manufacturers will pay high prices for quality grain.

5. Vegetables and fruits. Raising a few acres of truck crops or starting an orchard can pay large dividends, not only through numbers 1 and 2, above, but also through roadside fruit and vegetable stands and farmers markets. Community-supported agriculture (CSA) has become popular. Retail sales bring larger profits than wholesale. See other ideas in Chapter 19.

6. Check the advertisements in the publications that promote organic and ecologically sound agriculture, or run your own ad in these or in local farm papers or "shoppers." A list of organizations and resources is found in Appendix B. See Chapter 19 for additional ideas.

GET YOUR OWN SYSTEM

O
K, you say, I'm convinced that conventional agriculture has big problems and that some sort of ecological or natural system makes sense. I want to make basic changes . . . now, just tell me exactly what I need to do—how about a checklist: 1, 2, 3?

Unfortunately, it's not that simple. There is no magic formula for changing from financially shaky, chemical-intensive agriculture to an ecologically-sound, prosperous operation. There are so many variables and possibilities. Just in the United States, climates range from tropical in Hawaii to arctic in Alaska to desert in the West. Topography varies from table-top flat to rolling hills to mountainous. Soils can be fertile loam or pure sand or dense clay or alkali seep. Some regions are best suited for grains or row crops, while others are good for tree crops, or pasture or forages.

Then there are the interests, abilities, equipment, labor availability and financial possibilities of you, the farmer. *You need to develop your own farming system.*

Brainstorm. You should sit down (along with your spouse and any partners, financiers or managers) and assess your situation. Have a brainstorming session; write down your thoughts or tape record the session. Don't be afraid to dream, to imagine some pretty wild ideas. Think big. Imagine you could do anything; don't be limited by practicality at this stage. Think of goals. Try to fill needs or supply possible markets. Think of new ways to use a resource you already have—like corn cobs. Can corn cobs be made into toys—or paper—or alcohol? Combine elements from widely different sources.

Get additional ideas from what others are doing and from published resources, such as *Acres U.S.A.* and *Mother Earth News*. Watch for new and developing trends, such as public concerns about saturated fats, or growing biofuels.

Get real. After you have brainstormed for ideas, then it is time to be more realistic. What are your short-term and long-term goals and needs? Just for supplementary income or to start a whole new business? Do you want steady year-round income or will a seasonal operation meet your needs? Go through your ideas and pick out several that seem the most promising.

For each idea, take a sheet of paper and write the idea at the top, like **Raising Llamas**. Then write down why you want to do it, such as, "To provide a popular and unusual animal for an expanding market," or perhaps, "To develop a new market for llama milk."

Next, you really have to get practical. You need to do your homework and investigate the feasibility of your idea. Evaluate the pluses and minuses of your present situation. Find out what you will need to get started—and to grow bigger. Using our example:

- How much land do you need to raise llamas?
- Will llamas do well in your climate?
- What do llamas eat, anyway? Can you grow it?
- Do llamas need special housing?
- How many hours per day are required to care for a llama?
- How promising is the market for llamas? For llama milk?
- How much can you expect to pay for a llama?
- Etcetera, etcetera.

Naturally, you would like to make a profit right away, but many ventures will have sizeable start-up costs and/or may not get into the black for several years, such as growing tree fruit or pulpwood. Can you round up the needed capital—from loans, grants or investors? Check into the possibility of grants or matching funds from state or federal agencies or from private foundations. What is the likely cash-flow?

Check out the possible legal requirements or pitfalls, such as zoning laws, food safety laws, pollution control ordinances, and tax laws. Is insurance required?

Talk to potential markets to be sure they would buy your product and for how much. Do they have any quantity or quality requirements? Check out any transportation costs. Will you need to advertise; if so, which media are best and what will it cost?

Make out a detailed (as best you can) business plan and projected budget—for each year for five or ten years. If this is too much for you to handle, there are professional small business advisors available. Some government agencies provide their service at no cost.

Now, do you still want to try that idea? Is it a wise, practical idea? Think about it for a while. Discuss it with your family and others whose opinion you trust. Could your new venture survive an unexpected disaster such as illness or injury, or a devastating storm? Don't rush into anything. Probably it would be best to start small and see how things go. Don't risk everything at once. Maybe "Larry's Llama Milk" is an idea whose time has not yet come.

Suggestions. Perhaps you can merely fine-tune your current operation, cutting down on expensive inputs such as fertilizers, herbicides and pesticides; or becoming more efficient in other ways; or raising higher quality plants and/or animals. Or maybe you can expand your current operation into a related area that would require only small levels of new expenses, such as growing and packaging your own brand of popcorn, or processing your high quality milk into yogurt. These approaches should be safer and easier than jumping into an entirely new venture.

Following are some pointers from the book *How to Make $100,000 Farming 25 Acres* by Booker T. Whatley (1987):

1. Produce only what your customers want.

2. Produce high-value commodities (plants or animals) in such a way that different ones are ready to market at different times of year, thus providing year-round income and employment. Diversify into unusual products.

3. Be located on a hard-surfaced road within 40 miles of an urban center of at least 50,000 population.

4. Run a pick-your-own operation, so that much of the labor is done by the customer.

5. Have a guaranteed market with a clientele membership club with an annual membership fee (community-supported agriculture).

6. Be weatherproof, with drip and/or sprinkler irrigation.

7. Avoid middlemen. Market your products directly.

8. Be covered by at least $250,000 of liability insurance ($1 million is better).

Ideas. Here are some other ideas that you may find useful in increasing your profitability or starting a new business:

1. Buy used machinery, not new. Do your own repairs (maybe start an equipment repair business). Adapt machinery to new uses or invent new machinery (a possible new business).

2. Grow your own fertilizer in order to reduce purchased fertilizer inputs. Use green manures, legumes, animal manures and compost.

3. Grow your own high-quality seed (on contract or start your own seed business).

4. Use solar and wind energy, and grow biomass for alcohol or diesel fuels, or make methane fuel from manure.

5. Recycle wastes; make compost from a variety of organic matter, including city leaves, newspaper and wood waste. Start a compost business.

6. Do field research or run test plots for ag companies.

7. Add value to a commodity by doing your own processing, such as butchering meat, spinning and weaving wool, or making apple cider, maple syrup, sorghum, flour, baked goods, yogurt or ice cream, jams and jellies, pickles or salsa.

8. Raise high-quality crops and/or animals, commanding a higher market price. Become certified as an organic grower. Develop your own brand and logo, and begin intensive marketing.

9. Raise fish, turtles, frogs or crayfish for food or pet stores. Or raise fish bait.

10. Raise game animals and birds. Open a hunting preserve.

11. Raise unusual animals or rare breeds, which will command a high market price.

12. Raise goats for meat, milk and cheese.

13. Raise bees for honey and pollination services.

14. Raise veal or feeder livestock (humanely). Rehabilitate poor quality (cheap) livestock.

15. Raise high quality poultry for meat and "farm-fresh" eggs.

16. Grow high quality crops for forage (for yourself or to sell) or for making livestock feed pellets.

17. Grow unusual or high-value crops, such as:
   • Mushrooms (shiitake, maitake, oyster, etc.)
   • Herbs, spices, flavorings (such as mint oil)
   • Fresh flowers
   • Flowers for seed or to make dried flowers

- Wildflowers and prairie grasses for seed or bird seed
- Start tropical plants or house plants
- Watercress
- Specialty grains: amaranth, spelt, triticale, food-grade soybeans, flaxseed, quinoa, kamut
- Popcorn, especially unusual colors (red, blue)
- Vegetables, for early-season market, ethnic and specialty markets (possibly in greenhouses or with a hydroponic system)
- Fruit, melons, pumpkins, squash
- Old, heritage varieties of fruits and/or vegetables
- Sod
- Trees for nuts, maple syrup, lumber, pulpwood, fenceposts, firewood, seedlings for reforestation, Christmas trees, nursery stock

18. Start a tourist attraction, such as a campground, lake for fishing/swimming/boating, school tours, petting farm, riding stable, cornfield maze, bed and breakfast, or trails for hiking, skiing, snowmobiling or all-terrain vehicles. Have special "farm days" or a "harvest festival."

19. Provide farm work and a good environment for tourists, foreign-exchange students or disadvantaged youth (with possible funding from government agencies).

20. Start a pick-your-own operation with fruits, berries, vegetables.

21. Start a roadside stand. Sell at farmers' markets or flea markets.

22. Sell directly to grocery stores, restaurants, schools, health food stores or co-ops.

23. Start your own organic foods restaurant, health food store or organic foods co-op.

24. Start a membership community-supported agriculture (CSA) operation, with in-season produce supplied or delivered to customers each week. Keep in touch with customers through a website and/or newsletter.

25. Do custom plowing, planting, cultivating or harvesting.

26. Become an agricultural consultant and/or a dealer for products.

Entire system. It is good to figure out some new way to increase your income, or a new commodity to produce, but you also should be thinking about whether you should convert your entire farm to following one or another system of agriculture. Throughout this book, we have contrasted the problems caused by modern, conventional agriculture (over-use of toxic chemicals, degradation of soil and water, low-quality food, etc.) with the superior results of more natural methods, variously called ecological, biological or natural-system agriculture. Actually, there are

Will Biofuels Save Us?

The really big thing now is for the U.S. to break free from foreign oil by growing our own—*biofuels*. The best known are ethanol (ethyl alcohol, "grain alcohol") and biodiesel (mainly from waste vegetable oils and soybeans), Biofuels are one part of *biomass* production (obtaining useful products from living organisms); another is *biomaterials*, manufacturing such products as plastics, lubricants, solvents, fragrances, flavorings and various organic chemicals. Rubber, paper, cardboard and chip-board are well-known examples.

A major advantage of using biomass to make products is that they are sustainable, or renewable, and they grow by trapping free energy from the sun (photosynthesis), rather than "mining" the earth's limited supply of oil, gas and coal. One possible disadvantage is that harvesting and removing a major portion of the plants from the land will reduce the amount of organic matter that can be recycled into the soil, with possible depletion of soil fertility and structure.

But back to biofuels. The economics and politics of the world petroleum market have recently made biofuels attractive. The long gas lines of the 1970s and sky-high price increases after Hurricane Katrina have brought into focus the foolish wastefulness of our modern consumer-oriented society. Even in agriculture, many commercial fertilizers are made from fossil fuel. Until recently, biofuels were not economically practical, although they have been around for a long time. Rudolf Diesel, inventor of the diesel engine, first made it to run on peanut oil. Henry Ford designed early versions of the Model T to use ethanol. Then along came the growing petroleum industry and snuffed out those early biofuels.

The current interest in ethanol may not be the best solution. So far, most commercial ethanol has been made from the fermentation by yeast enzymes of carbohydrates (starch, sugar) in corn (in the U.S.), flaxseed and rapeseed (in Europe), or sugar cane (mainly in Brazil, which has become completely self-sufficient in fuel). The left-over corn mash can be used for livestock feed. It was once believed that corn-based ethanol production used more energy than the burned fuel produced, but that is not true; still this type of biofuel process is not very efficient as far as alcohol produced per unit of biomass. Also, the use of crops grown with fertilizers and a lot of water on prime agricultural land for fuel production is not the best land use. Researchers are working to develop more efficient fermentation using cellulose besides sugar and starch (partly with different microbial enzymes, or else by genetically modifying yeast), and to use biomass from wastes (manure, crop stalks and straw, rice husks, forest product wastes, cannery wastes, city leaves or brush, sewage and garbage) or plants that grow well on poor or droughty soils, such as fast-growing poplar or willow trees, drought-resistant shrubs and weeds (mesquite, sumac, milkweed, goldenrod, thistle, sunflowers, spurge, etc.), as well as various exotic and prairie grasses (especially switchgrass, which can produce several tons/acre of biomass, with one ton producing up to 96 gallons of ethanol). One wonders about the wisdom of using fragile desert and prairie lands for growing crops . . . remember the Dust Bowl?

Biodiesel fuel can be made from soybean or sunflower seed oils, as well as just about any other vegetable oil, either freshly-produced or waste cooking oil from restaurants. Certain algae are the most productive biomass source.

Are biofuels alone the answer to future petroleum shortages? Many analysts believe we have now peaked at about the half-way point in exploiting the earth's petroleum reserves, with the remaining half much harder to find and extract than the first half. With the U.S. by far the largest user of oil, and far too wasteful, future development of biofuels will become at least a partial (and temporary) stopgap, but the day of reckoning can be put off far into the future if we just have the political and individual will to conserve the current supply (drive more fuel-efficient vehicles, use mass-transit, eliminate unnecessary travel and other energy use), along with further development of solar and wind energy. Nuclear energy is not a good alternative; there is currently no easy way to deal with the radioactive wastes.

several philosophies or types of natural-oriented agriculture. The more prominent ones are:

1. *Organic agriculture.* This is the most commonly recognized natural system of agriculture, and we covered it in Chapter 15. Briefly, organic agriculture as an organized system was originated in the early 20th century in India by Sir Albert Howard, who developed a method of composting agricultural and city wastes. After finding that his compost helped grow healthy, productive crops, he published books espousing proper care of the soil, especially recycling organic wastes and fostering beneficial soil organisms. In the U.S., J.I. Rodale further refined and demonstrated the effectiveness of the organic method and published numerous books and magazines. In the late 20th century, organic agriculture became formalized by a set of rules and is regulated by certifying organizations, and eventually by the USDA.

The basic philosophy of organic agriculture is that fertile soil is maintained by replenishing it with compost rather than chemical fertilizers. The use of commercial fertilizers and highly toxic pesticides should be avoided. There are different degrees of "organic" farmers. The purists absolutely refuse to use any synthetic fertilizers or pesticides, while many others, although they do not qualify for certification, may use small amounts of some less harmful fertilizers and/or small doses of herbicides or insecticides if needed. Using antibiotics and growth-promoting hormones on animals is forbidden, and large confinement animal operations violate the spirit of organic agriculture.

For the most part, organic farmers rely on recycled organic matter and natural rock fertilizers for soil fertility, on cultivation or other non-chemical methods for weed control, and on natural resistance and natural enemies for pest control.

One problem some farmers have is transitioning from chemical-intensive farming to organic. Quitting chemicals "cold turkey" could result in a disaster. Abused soil may take several years to recover. Current certification organizations allow three years for transition and provide considerable advice and guidance. There are now many federal, state, university and local-farmer organizations doing research and providing useful information. As we have mentioned previously, organic methods can produce good to above-average yields of healthy, nutritious food.

2. *Biodynamic agriculture.* A somewhat different system of soil stewardship is called biodynamics. Much of what is done is similar to organic agriculture, but biodynamics has a deeper philosophy behind it.

Biodynamic agriculture began when Rudolph Steiner, a German philosopher, gave a series of lectures to a group of farmers in 1924. Steiner had developed a philosophical view of life called anthroposophy ("wisdom of man"), based on the ideas of the well-known German philosopher Johann von Goethe. Anthroposophy is a "path of knowledge that strives to lead the spiritual in man to the spiritual in the universe," according to Steiner. Living organisms are believed to have a spiritual nature. The bio-dynamic system of agriculture encompasses the farm as a whole, aiming toward a "farm organism" in harmony with its habitat, its workers, and the market it supplies.

Dr. Ehrenfried E. Pfeiffer studied Steiner's methods and transplanted them to the United States in 1939, on a farm in Spring Valley, New York. He started the Bio-Dynamic Association and developed a bacterial compost starter inoculant.

Bio-dynamic principles are rather rigid in what they require the farmer to do. The emphasis is on restoring the soil's humus through supplying specially-made compost. The compost must be made in an exact way from animal (cattle) manure and other organic matter, soil and small amounts of special plant "preparations" (made by burying certain wild herbs along with specified animal organs for a summer or a year). The fermented preparations are inserted into a compost pile, where they are said to influence the composting process to proceed in the proper direction. Some preparations are also used as either foliar or soil sprays.

Also, the farmer must use a crop rotation, which alternates soil-exhausting with soil-building crops. Companion planting is encouraged, and a best ratio of pasture and cropland, and a varied environment surrounding the farm are the ideal goal (a harmonious distribution of fields, pasture, forest, lakes and rivers).

Tests comparing bio-dynamic methods with conventional agriculture have shown crop yields to be comparable, and quality and disease resistance to be excellent. In an animal-feeding test, chickens were either fed grain from bio-dynamically fertilized soil or from conventionally fertilized soil. Thirty eggs from each group of chickens were stored. After two months, four (14%) of the conventional eggs had spoiled, compared to only one (4%) of the bio-dynamic eggs. After six months, 18 (60%) conventional and eight (27%) bio-dynamic eggs had spoiled.

Although it is fairly popular in Europe, biodynamics does not have many followers in the U.S.

3. *Reams and the RBTI*. If anyone could cause bewilderment and head-scratching among agronomists and soil scientists, it was Dr. Carey Reams. He developed a system of soil fertilization that contradicts many of the establishment's beliefs. But it is based on research from the 1920s and '30s, and it generally gives excellent results. Some of Reams' claimed yields included 20 tons of alfalfa per acre, 100 bushels of soybeans per acre, 90 bushels of wheat, 20 tons of cabbage, 20,000 quarts of strawberries, and so on.

Reams was originally an engineer, and in the course of solving problems for clients, he became interested in health and agriculture. Through reading the works of early researchers and advising farmers and horticulturists in the southeastern U.S. and elsewhere, he developed what he called the Reams Theory of Biological Ionization (RBTI). This theory contains some ideas that conflict with chemistry and agricultural textbooks and is difficult for someone with a background in "standard science" to understand, because Reams defined many terms differently than scientists do. Reams also applied his theory to animal and human health and taught his principles and methods, along with his student, veterinarian Dr. Dan Skow, in seminars and books. Reams died in 1985, but Skow and his associates continue his work.

Some of the principles of the RBTI are as follows. The theory is based on the idea that all occurrences involve some type of magnetic energy or electrical charge. Magnetic or electrical energy from the earth's surface is what makes plants grow. The electrical potential of the soil can be measured by a conductivity meter, in units Reams named "ergs" (equal to the standard scientific units called micromhos). Some nutrient ions are cations (positively charged) and others are anions (negatively charged). Reactions ("resistance") between anions and cations in the soil produce energy for plant growth. Anionic elements are calcium (contrary to standard chemistry), potassium, chlorine, sodium, nitrate nitrogen and sometimes oxygen and hydrogen. All other elements are cations, including ammonium nitrogen.

A preponderance of anionic soil nutrients causes plants to produce vegetative growth (stems and leaves), while cationic elements cause flowering and seed production. Thus, by regulating the balance of soil nutrients, it is possible to control the time of flower-set of a "seed crop," or to keep a "leaf crop" from going to seed (such as hay or lettuce).

Most of the practices that Reams recommended are compatible with organic methods (such as recycling organic matter, fostering soil organisms, and the use of some rock fertilizers), but there are some differences. He recommended using some of the synthetic "chemical" fertilizers, but only ones that are compatible with soil life and in smaller amounts. He didn't hesitate to recommend using certain insecticides in low dosages, but did not like herbicides. He made use of foliar feeding to help a crop along and emphasized crop quality as measured by sugar content with a refractometer.

4. *French intensive method.* This is a method of intensive production from a small area of land that has been introduced from Europe into California. It requires much hand labor, so it is not economical for large commercial operations, but can produce excellent yields of vegetables for family-sized farms. Crops are grown on raised beds, about 154 square feet in size, which have been deeply tilled (24 inches) and highly fertilized. The beds are surrounded by fresh horse manure to protect from spring frosts. Yields run from three to five times greater than with conventional methods.

Or make up your own. Quite possibly, none of the above natural-oriented systems of agriculture will appeal to you. They may be impractical for your needs, or too complex, or maybe you don't agree with their philosophy. Fine, just put together your own system. Using the principles and laws of nature—how the soil, plants and animals work, along with the interrelationships of ecology—pick and choose your methods and practices to best accomplish your goals—in an ecologically sound way as much as possible.

A New Old Approach

Recently a group of USDA agronomists and computer-oriented specialists have developed a "new" approach for more flexibility and better efficiency, called *dynamic cropping systems* (*Agronomy Journal*, 2002, Vol. 94, p. 957–961). Noting that a number of unpredictable factors can negatively affect a farming system, such as weather, fluctuating markets and new technology, the authors take a holistic view and recommend getting many kinds of information (soil tests, possible crops that can do well in your climate, weather forecasts, market trends, etc.) and developing your own *flexible* cropping system, one which is based on a large diversity of crops and management techniques and which can adapt to changing conditions and opportunities.

Great emphasis is placed on a long-term, sustainable outlook that not only maximizes soil productivity and profits, but also one that is socially acceptable and environmentally sound. Crop rotations are used when and if they make sense, and the exact sequence of crops depends on changing weather, soil and market conditions, rather than sticking to a fixed "cookbook" rotation. Reduced input costs and a multiple-enterprise system are important.

It is certainly refreshing to see the "academic agriculturists" engaging in such far-reaching, common-sense thinking. Although their ideas are partly new, probably generations of wise farmers have used flexible systems.

Do whatever you do for a good reason. Don't spray toxic chemicals . . . just in case. Don't get out and plow just because your neighbors are. Know what each operation or input is supposed to do. Does it improve or degrade your soil's structure or earthworm population? Will it increase or retard crop growth? Will natural predators and parasites of crop pests be killed? It takes some basic knowledge . . . and lots of common sense!

Again, start small and see how it goes. Maybe convert one field each year to your new system. Since abused soil takes a few years to recover, a slow transition should keep your income more or less constant.

Keep a positive frame of mind, knowing that natural systems have been running the earth for millions of years. There could be temporary setbacks, perhaps from bad weather, but if you don't give up, you will achieve good results.

Look to the future. Above all, your new system should be sustainable— able to continue successfully, indefinitely. With rising input costs— from fertilizers to seed to fuel to transportation—plus the increasing environmental damage caused by conventional agriculture, the odds are definitely swinging in favor of natural-system agriculture. Conventional energy-intensive agriculture is not sustainable. With world conditions as they are, modern agriculture cannot last. It seems so obvious and logical:

if you rely on fertile soil to grow your food, then doing anything that destroys or pollutes soil can't continue . . . if you want to eat.

After astronauts visited the moon and looked back at the beautiful blue and white "marble" from which they came, the earth was called a fragile "spaceship earth." It is really sad that today, over 35 years later, the only place in the universe able to support us is in even more peril than then. Besides the destruction of natural habitats by development and lumbering, now the entire oceans are seriously overfished, and more urgent, global warming promises future ecological and social disruption. Will our leaders make wise decisions?

Land use. So, with a sense of ecology in mind, what are the best choices in how we use the limited supply of land on earth? Efficient and proper land use is only a matter of common sense. The earth's surface differs from place to place in topography, soil type, climate and native vegetation. It only makes sense to use land for purposes that are in harmony with the local topography, climate and natural resources.

But in all too many cases, that has not happened. The lessons of the dust-bowl 1930s have not been remembered. The tree/shrub wind-slowing shelter belts that were planted in the plains after the dust bowl days have been ripped out to make room for huge irrigation rigs . . . which are sucking dry the underground aquifers . . . all for short-term profits from crops that never should be grown there. Erosion-prone row crops are grown on hillsides, with no cover crop over the winter. Woods and fence rows are torn out, eliminating bug-eating wildlife. Suburbs and freeways sprawl out over prime farmland. Just between 1992 and 2002, 2.2 million acres of land were lost to development.

Best agricultural land use should suit the local conditions and resources. Uplands and sloping land should be used for pasture and hay crops, with row crops on flatter lowlands. Rows should be tilled around the contour, not up and down a slope. Crops using high soil moisture should not be grown in low rainfall areas, Land that will only support a forest should be left in forest; wild trees can be profitably harvested for wood or paper, or a nut or fruit orchard can be planted. Housing developments belong on hillsides or wooded land, not on prime agricultural land. The accompanying chart gives common-sense land uses for temperate climates.

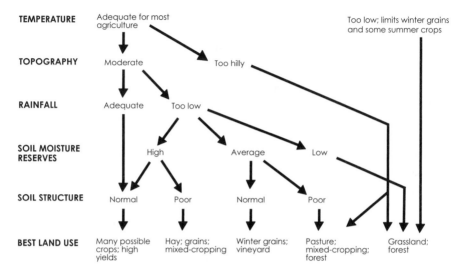

TEMPERATURE	Adequate for most agriculture				Too low; limits winter grains and some summer crops
TOPOGRAPHY	Moderate		Too hilly		
RAINFALL	Adequate	Too low			
SOIL MOISTURE RESERVES		High	Average	Low	
SOIL STRUCTURE	Normal	Poor	Normal	Poor	
BEST LAND USE	Many possible crops; high yields	Hay; grains; mixed-cropping	Winter grains; vineyard	Pasture; mixed-cropping; forest	Grassland; forest

Intelligent land use (adapted from R. Blanchet, 1974, in Fertilizer, Crop Quality and Economy, *V.H. Fernandez, ed., p 367).*

Make your farm an ecosystem. It certainly isn't possible in many cases, but the ideal environment on and around a farm is a wide variety of habitats, with hundreds of species of plants and animals. As we saw in Chapter 2, the multitude of plants, animals and microbes in any habitat interact in complex ways, depending on one another for food and other things. A bird may eat a frog which eats an insect which eats a plant which relies on soil microbes for many nutrients . . . and then a raccoon may eat the bird, and a bobcat may eat the raccoon and most of them probably have parasitic lice or mites! Altogether, it's an ecosystem.

Ecologists have found that especially in large, complex ecosystems, the many species all interacting and depending on one another give a *stability*, or resistance to change, to the system. That is, if one species (such as an insect) becomes more numerous, more of them will be eaten by frogs, and then if frogs multiply excessively, more of them will be eaten by their predators. This is one of the values of *biological diversity*, a wide variety of species in any one habitat.

Along comes one species called *Homo sapiens* and everything goes haywire. Motivated by arrogance and greed, humans have spent centuries seeking to overpower and control nature, and in these days of burgeoning population, man's effects on the natural environment have been too often devastating. We have cited one example after another throughout this book. Ecosystems and species diversity are threatened all over the earth.

By its nature, agriculture is to some degree artificial. Plant or animal species not native to the area are raised, and usually the original vegetation is disturbed or eliminated as fields or pastures are created. The worst example of unnatural agriculture might be irrigated fields, several hundred acres in size, of monoculture corn or soybeans, side-by-side with others, blanketing the land from horizon to horizon, drenched with herbicide and insecticide! There is no place for the insects or birds that might eat those weeds and pests, so the poor farmer spends thousands of dollars for the toxic chemicals. Remember, when a natural part of an ecosystem is eliminated, the farmer has to supply a substitute.

Ideally, a farm ecosystem should include a variety of crops and animals, perhaps small grains and corn or sorghum (interplanted with a ground cover), along with legume and grass forages, and pastures containing a mixture of native grasses and wildflowers. Several kinds of animals would be great, perhaps beef or dairy cattle, pigs, sheep or goats, and poultry, grazing in the pastures, of course, not in constant confinement. Between neighboring fields and pastures, the fence rows would be several feet wide and filled with grasses and wildflowers. Interspersed among the agricultural land would be woods or brush, wetlands and possibly a stream or lake. In all, this would provide habitats for hundreds and hundreds of wild species, few if any of which would be pests or problems. One of the few scientific studies on the effects of agriculture on wildlife (*Science*, 2005, Vol. 307, p. 550–555), states that even low-intensity agriculture eliminates half or more of the species normally present in an area.

Depending on your land holdings and the natural resources of your part of the country, many farms cannot easily include all of the above features, but still, trees and native pastures can be planted and lakes can be built. The most important factor is the value system of the farmer. You must appreciate the services rendered by, and the just plain beauty of, nature's marvelous organisms. You need to be able to sacrifice possible profits from part of your land, realizing that the free services of a diverse, healthy ecosystem may well give you a better bottom line in the end. It may take some courage to let go of the profit-seeking, control-nature mindset, but you can do it. Be a steward, not a plunderer.

Chapter 20

FINE TUNING

I t is human nature to want a short-cut or easy way to accomplish a task, such as the proverbial "silver bullet" and "magic wand." Wouldn't it be great to sprinkle some "foo-foo dust" on your fields and in a few days have your soil turn mellow and your crops grow a foot?

Unfortunately, it probably won't work that way. There is no simple "cook-book" magic formula or product that will absolutely cure sick soil and the plant and animal problems it causes.

Assuming that your soil has some major problems that need fixing, the best approach (and cheapest in the long run) is to work on the basics first, and then do some fine tuning. It is most important that your soil has good structure (loose, crumbly), is well-aerated and well-drained (no hardpan or waterlogging), with plenty of humus and beneficial soil organisms, and with a proper balance of plant nutrients.

Review especially Chapters 3, 6, 14 and 17 concerning soil and the many ways to restore and improve it. There are tillage and cropping methods, and the addition of rock-based soil amendments and organic matter that should improve soil structure, aeration and drainage. And there are many kinds of liming materials, fertilizers and incorporated crop residues that can supply nutrients. Number one is good soil structure, and number two is correcting the balance of major and secondary nutrients (calcium, nitrogen, phosphorus, potassium, magnesium, sulfur). Only then should you worry much about fine tuning with the minor or trace elements, or improving crop growth with special plant stimulant products. Except when a trace element is extremely deficient or excessively abundant (and possibly toxic), in most decent soils, when the major and secondary elements are in balance, the trace elements will be in fairly good balance also.

So, if your basic soil corrective measures are well underway, and there are clear signs of improvement, then you can look at possibly trying some fine tuning, especially some ways that are promoted to improve crop growth and performance. Some of these fine-tuning methods are well-grounded in scientific explanation and have been tested and proved, while others seem pretty "far-out." I make no claims or recommendations for them, but merely report about them.

Better crop growth. There are a number of exciting methods which can increase the growth, health, yield and quality of crops. They include seed treatments, growth stimulants, foliar feeding, microbial "helpers," natural pest and weed control, efficient cropping systems, effective fertilizing materials, and the use of weak natural energies.

1. *Seed treatments.* There are various kinds of seed treatments with different purposes. Some are coatings of fungicides or insecticides to prevent soil-borne diseases or pests from attacking. Although such poisons should not be needed in healthy, fertile soil, they can be effective in less than ideal soil and under unexpected stress conditions (waterlogging, cool weather); also, this approach is good because smaller amounts of poisons are used than if all the soil were treated.

Inoculation of seeds with beneficial microorganisms has been done for years in the form of root nodule bacteria (*Rhizobium*) for legumes, with excellent success. Inoculation with general beneficial root bacteria (rhizosphere bacteria) and fungi (mycorrhizae and nematode-trapping fungi) is still somewhat experimental, but is being developed commercially and will definitely be important in the future. These microorganisms can help roots fight harmful microbes and nematodes, and help roots absorb soil nutrients (see Chapter 2).

Then there are commercial seed treatments that contain growth-stimulating enzymes, hormones, and trace elements. These get the germinating plant off to a faster, more vigorous start and stimulate a bigger root system, which are important in cool or northern climates. Tests at Clemson University found that soaking beet seeds in seaweed extract for 30 minutes increased germination 25%.

2. *Growth stimulants and regulators.* Plant growth regulators are either natural or synthetic substances that act in small amounts to affect plant functions so that growth is increased (sometimes flowering or fruit drop

is delayed), quality is improved, yield is increased, or harvest is facilitated. Natural growth regulators include plant hormones and seaweed extracts. They often result in increased yield, quality, disease and pest resistance, and cold hardiness, but if the soil and plants are already healthy, there may be little effect.

There are also some products called activated water or catalyst altered water, made by treating water with electricity or an extract of lignite coal. The molecular configuration of the water is said to be changed, making it more active. When such treated water (diluted with ordinary water) is applied to soil, seeds, or foliage, plant growth is stimulated. It can also be added to livestock feed and used for human health.

Some farmers and gardeners add a small amount of hydrogen peroxide to water which is used to irrigate or spray on crops. The hydrogen peroxide supplies oxygen. It can also be used in the drinking water of animals or humans to improve health or prevent disease.

Various synthetic growth regulators act as herbicides for some species and growth stimulants for others. Too much of a growth regulator, or applying it at the wrong time of plant development or in the wrong weather, can cause growth inhibition or plant damage. Some products increase yields of some crops, but not others. Synthetic chemicals may be toxic to fish or wildlife.

3. *Foliar feeding*. We have previously mentioned briefly that applying materials to leaves, or foliar feeding, can be used to supply small amounts of nutrients, mainly trace elements. Also, we mentioned that a crop can be induced to "switch over" from vegetative growth to begin flowering and seed production by the right foliar spray. Another use of foliar spraying is to apply some of the growth regulators covered in the previous section. Since they are needed in only small amounts and at certain times, foliar spraying is the best method of application. Such sprays are often used on fruit trees to increase blossom set, and to improve shipping and keeping quality, fruit color, and ripening.

4. *Microbial helpers*. Besides the use of microbial inoculants as seed treatments, similar inoculants have been developed for general soil application. Nitrogen-fixing bacteria and algae, which do not have to live in root nodules, are one approach; they can benefit non-legume crops. A general population of beneficial soil bacteria or the special rhizosphere species is another approach. A third is the special symbiotic fungi, mycorrhizae. If these beneficial microorganisms become successfully established in the soil, they should remain active for years (no toxic

chemicals should be added, of course). It is often difficult to get the microbes successfully established, however, because microbes already in the soil may "out-compete" them. An excellent inoculant of beneficial microbes is compost; another is poultry manure. Still another is using water that has had compost, animal manure, or plant residues (old hay, weeds, or grass clippings, etc.) soaking in it. Certain special soil-inhabiting algae can also improve growth and yield.

5. *Natural pest, disease, and weed control.* We have previously discussed allelopathy (Chapter 16), the ability of the secretions or residues of one plant species to inhibit or kill another. Some weeds are allelopathic, quackgrass for example, and can seriously stunt crops. Certain crops or their residues are known to inhibit other plants, and the proper management of these "natural herbicides" can reduce or eliminate the need for ordinary herbicides.

And of course, the old practices of controlling weeds by timely cultivation or frequent mowing are very effective. Cultivation has the added benefit of breaking a crust and aerating the soil, which can greatly improve crop growth.

Healthy plants have built-in mechanisms to resist diseases and pests. We have pointed out how root-inhabiting bacteria and fungi protect against soil-borne diseases. Recent research has shown that plants produce various chemicals that repel or kill invading pathogens or pests. Some of these protective chemicals are only produced when microbes start to invade or pests attack

The concept of integrated pest management (IPM), is the right approach toward short-term pest control problems (the best long-term approach is building up healthy soil so plants will be healthy and naturally resist pests, and encouraging natural enemies and beneficial soil organisms). In integrated pest management, the idea is not to completely eradicate a pest species, but to keep its population levels below the point of doing serious damage. This is done by using non-toxic controls when possible (natural enemies, management practices) and toxic chemicals only when necessary.

6. *Efficient cropping systems.* The method of multiple cropping and intercropping have already been covered in Chapter 14. They can greatly increase crop productivity per acre.

7. *Effective fertilizers.* As an alternative to using large amounts of synthetic commercial fertilizers, which are often expensive or may harm soil organisms or soil structure, there are several fertilizer materials and

Injecting liquid fertilizer as a split application.

methods to consider. Some of the soil conditioners we covered previously also have fertilizer value. These include liming materials, rock phosphates, other ground rock trace element fertilizers, humates, seaweed, compost, and other organic matter. Applications of finely ground glacial gravel (or gravel crusher screenings) provides a "shot in the arm" of major and trace elements, giving terrific crops: two to four times increase in yield of garden produce in California (including 18-foot pole beans), 12-foot clover in

Vermont, and a 15 bushel/acre increase of corn in Michigan. You can make an effective liquid fertilizer for nothing by mixing organic matter (such as manure, old hay, garbage) in a stagnant pond or water tank and waiting until the "brew" is good and ripe. The decomposing organic matter will leach nutrients into the water, and if nitrogen-fixing bacteria are active, you can get a great increase in nitrogen content. Dilute and strain the mixture before spraying on your land. Adding some compost to the mixture would be an excellent way to introduce beneficial microorganisms. A compost-based extract is often called "compost tea."

As we have talked about, careful use of commercial "N-P-K" fertilizers can give good results, provided the right materials are chosen. Materials that harm soil structure (anhydrous ammonia) or harm soil organisms or seedlings (anhydrous ammonia, chloride-containing materials, high-salt materials) should be avoided. Which materials should be used and the amount depends on many factors: the crop, time of year, past fertilizing history, method of fertilizing (broadcast or row), and soil needs shown by soil tests. Generally, if possible, commercial N-P-K materials should be used in smaller amounts to supplement natural fertilizers or for special uses, such as side dressing in mid-season to supply needed nutrients or to "switch" a crop from vegetative growth to seed production.

If the soil is clearly deficient in one or more trace elements, application of the right materials can make a big difference in yield and quality. The potential for yields in really good soil is amazing. On test plots and fields in Idaho and Washington, open pollinated corn has yielded over 400 bushels per acre, and dry-land wheat over 250 bushels per acre (with 19.6% protein), according to Bill Johnson of Soil Spray Aid, Inc., Moses Lake, Washington, in a 1983 talk.

8. *Natural energies.* There is a whole world of energy surrounding us and influencing plant, animal, and human growth and well-being. We literally live in a sea of energy. Besides the obvious light energy we receive from the sun, there are also invisible ultraviolet and infrared light wavelengths, and natural and man-made radio and microwave, X-rays, and gamma rays. These are called electromagnetic radiation, and they occur in different wavelengths or frequencies (see diagram). Everything in the universe— stars, rocks, plants, and ourselves—radiates electromagnetic waves at a frequency depending on its temperature, which is due to the vibrations of the atoms or molecules of which the object is composed. Very hot objects radiate infrared, visible light, ultraviolet, X-rays, or gamma rays. Cooler objects give off infrared, microwaves, or radio waves.

The Electromagnetic Spectrum

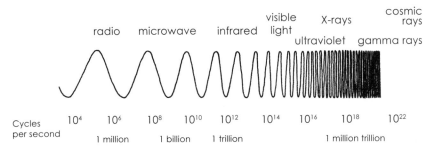

visible
light

cosmic
rays

radio   microwave   infrared   X-rays

ultraviolet   gamma rays

Cycles 10^4 10^6 10^8 10^{10} 10^{12} 10^{14} 10^{16} 10^{18} 10^{22}
per second   1 million   1 billion   1 trillion   1 million trillion

Besides these energies, there is electrical energy: static electricity, lightning, DC and AC. Electrical charges possess a force field around them. Fluctuating electric charges can be used to generate electromagnetic radiation; this is how a radio station broadcasts.

The earth has a magnetic field. All objects on the earth are subjected to it. A change or movement of an electrical charge generates a magnetic field, and a change in a magnetic field generates an electrical field. Charged particles (electrons, protons, ions) in motion are affected by a magnetic field, and charged particles themselves can create an electrical field that affects living things. It is now known that an excess of positive ions (common in modern buildings and certain dry south winds) at first causes people to feel energetic, but after a while they "overload" the body's metabolic processes and hormone system, leading to anxiety, migraine headaches, arthritis flare-up, asthma, high blood pressure, higher accident and suicide rates, etc. Negative ions (common in mountains, forests, and near waterfalls) or a "natural" balance of five positive to four negative ions, cause a sense of calmness and tranquility, sound sleep, and help heal burns, asthma, etc. Plants grow unusually large with positive ions.

Cosmic rays (nuclei of atoms stripped of their electrons) bombard the earth from outer space, along with other charged particles; protons and electrons from the sun (the solar wind). The beautiful aurora borealis (northern lights) results when these particles interact with the earth's magnetic field and atmosphere. The sun also has a magnetic field which is variable and reaches and influences the earth's field.

Living cells, and the tissues and organs they compose, generate weak electrical and magnetic fields. Ions (electrically charged molecules) can be held or "pumped" to one side of a cell membrane, creating an electrical charge. This is the origin of the electrical impulses in animal and human muscle and nerve cells. Plant cells also generate electrical currents. The inside of pea root cells is about 110 millivolts compared to the outside, while oat root cells are about 84 millivolts. The microscopic threads of a fungus produce currents that enter at the tips and exit farther back. Even the various membranes within cells are charged because of unequal sodium, potassium, and chloride ions and act as microscopic capacitors. Evidence indicates that the self-generated electric fields of a growing plant help control its development by distributing growth hormones.

Weak electrical currents and electromagnetic fields are known to affect cell activity. The right kind of current can speed up cell functions. Microwaves have been found to be absorbed by DNA, the material that makes up genes, with possible damage and cancer initiation. Weak electromagnetic fields can alter the flow of calcium ions in brain cells, cause animals' body temperatures to rise, increase the number of white blood cells, and slow down learning. Scientists are concerned about adverse effects on people living near radio broadcasting towers, large power lines, and other sources of extremely low frequency (ELF) radiation. Even common household electrical appliances and wiring are believed by some to aggravate asthma, high blood pressure, headaches and other ailments in those people who are sensitive to magnetic fields. The effects of stray voltage on dairy cattle may have a similar cause.

Sending low-voltage pulsating electrical currents into the body through electrodes on the skin has been able to control the pain of neuralgia, headache, childbirth, and arthritis, as well as helping circulatory problems, speeding the healing of bones and burns, and improving memory.

Way back in the early years of the 20th century, experimenters in England treated seeds with weak electric current for several hours (seeds were soaked in a weak brine to carry electricity). Results included an increase of 6 to 19 bushels/acre of oats, 6 to 12 bushels/acre of wheat, and 21 to 50% increase of barley. More recent experiments along the same line have been done, with promising results. (*Acres U.S.A.*, September 1981, p. 4).

Recently, iron particles (magnetite) have been found in the cells of certain kinds of bacteria and algae; and in the heads of many animals, including fish, sea turtles, birds, rodents, monkeys, and humans. This may be how birds and water creatures orient themselves during migration.

Living things are affected by external magnetic fields. Russian scientists have found that seeds placed in a moderate magnetic field for 10 days germinate faster than untreated seeds, and the seedlings grow faster (for strong fields, a shorter exposure time is sufficient). For example, in four days, magnetically treated barley roots grew 34.72 mm long and the shoots 21.10 mm, compared to 29.35 mm and 15.51 mm for untreated plants. Treated soybean roots grew 22 mm and the shoots 88 mm, compared to 6 mm and 68 mm for untreated plants. Results are better if the long axis of the seed is parallel to magnetic lines of force. Some crops mature earlier and produce higher yields when magnetically treated. The cells of magnetically treated plants apparently function more efficiently, producing more sugar but using less energy to metabolize. More dividing cells are present in the growing tips of roots and stems. Seedling roots grow out parallel to magnetic lines of force, and grow away from a stronger magnetic field and toward a weaker one. Canadian research found that some plants' roots will mainly grow outward in a north-south plane (winter wheats, some spring wheats, wild oats, some weeds), while others do not (barley, flax, corn, sunflowers). If crops whose roots do grow north-south are planted in rows running east-west, they will more effectively use between-the-row soil nutrients and water (*Crops & Soils*, 1968, Vol. 20, No. 8, p. 8–10).

Some researchers claim that the south pole of a magnet increases growth and that the north pole decreases it (the south pole is defined as the pole that turns toward the earth's North Pole, since opposite poles attract). In one experiment, crops grown from seed treated with the south pole of a magnet had better growth, yield, protein, sugars, and oils than those treated with the north pole. Laboratory rats living in south-pole magnetism were larger and healthier than those living in north-pole magnetism, although they died sooner, as reported in *Magnetism and Its Effects on the Living System*, 1974. Some people claim to have been cured of various ailments by wearing small magnets next to their bodies, including arthritis and cancer. Weak magnetic fields, such as the earth's magnetism, are known to influence cell growth, life span, and activity of organisms.

Different molecules or substances have a "natural frequency," and they absorb electromagnetic waves much more readily at that frequency than others, according to *Science News*, 1984, Vol. 125, p. 248. A number of people speak of cells or whole plant or animal bodies as having a certain frequency. In the 1920s the Russian engineer, Georges Lakhovsky theorized that the nucleus of a cell is like an electrical oscillating circuit, emitting and receiving electromagnetic vibrations at a certain frequency if the cell is healthy. If the cell was unhealthy or dying, there was a lower frequency. He built machines emitting short wavelength energy and was able to cure plant cancers, make plants and animals grow faster, and cure hospitalized patients of cancer, radium burns, goiter, and other "incurable" ailments. The best results were obtained with a machine called a multiple wave oscillator, which allowed every cell in the body to vibrate at its own best frequency.

Agricultural applications. In the mid- and late-20th century, devices based on the multiple wave oscillator were developed for agriculture. Jerry W. Fridenstine invented what he called an orthomolecular multiwave oscillating generator (often called a tower), which is said to "broadcast" certain frequencies (the individual frequencies of each crop and animal species) over a farmer's land. Some impressive results have been obtained, such as alfalfa growing 48 to 56 inches high and yielding over five tons per acre on the first cutting, full silos on much less acreage, over 200 bushel per acre corn in Colorado, more digestible animal feed, and contented, healthy dairy cattle, with an increase in milk production. The towers are said to focus and broadcast natural energy found in the earth's surface, to sort-of "perform acupuncture on the earth" in order to balance out too-high or too-low elements in the soil or animal feed.

A somewhat different device has been invented by Dr. T. Galen Hieronymus and later improved upon by Hugh Lovel, consisting of a vertical pipe about 10 feet long, with the bottom 2 feet buried in the ground. A slanted collector plate at the top of the "cosmic pipe" is said to collect "cosmic" or "eloptic" energy and transmit it into the soil through a radiating coil at the bottom. It is said to cause plants to grow faster, be healthier, and produce larger, higher quality yields within a radius of about one-half to over one mile. Results, which may take six months, include five-foot high alfalfa (double the previous year's yield in one cutting) and five-foot marigolds.

Supposedly, electromagnetic energy flows from the equator to the poles through organic-matter-rich soil. This energy helps plants to grow better. Some research indicates that there are north-south lines of greater energy (called ley lines), and that trees and other vegetation grow larger along these lines. The energy also goes in or comes out of the earth's surface at certain places, called vortexes. Man's activities and structures, such as roads, pipelines and power lines, interrupt the normal flow of "earth energy." Increasing the humus content of your soil will increase the energy available to plants. The commercial salt fertilizers of modern agriculture make plants grow fairly well because salt is a good electrical conductor, increasing energy flow. But at a price. Soil bacteria and fungi are disturbed or killed, humus is depleted, and the soil eventually becomes "dead," sterilized.

One of Jerry Fridenstine's towers. Parts of the tubing are filled with special mixtures of minerals.

The same type of thing can be done on a small scale, such as in your garden or flower beds. There are various ways of concentrating or focusing the "earth energy." In 1925, Georges Lakhovsky set up a coil of copper wire (one foot in diameter) around a geranium plant. The plant grew—and grew—and grew. By 1928, it was four and one-half feet tall. Another method is to stick two 18-inch copper wires or rods or large nails into the soil, north and south of a plant. Or simply sprinkle iron filings on the soil. Growing plants on raised beds is also supposed to increase energy.

The source of this plant-growing energy is the sun. Besides the light and other forms of energy emitted by the sun, the weak electromagnetic energy of the soil streams out from the sun. Dr. Philip Callahan, a researcher well known for his work on how insect pests locate the plants they attack (see Chapter 16), has expounded a theory on how it works. In his book, *Ancient Mysteries, Modern Visions* (1984, p. 104–109), Dr. Callahan said that the energy from the sun is a combination of magnetic and electrical charges called magnetoelectric monopoles. A monopole is like one end of a magnet. There are two kinds, a north magnetic pole combined with a positive electrical charge, and a south pole with a negative charge. Both kinds stream from the sun and strike the earth. The north monopoles are mainly absorbed by plants and move down into their roots. The south monopoles are absorbed by many kinds of rocks and soil, and they can flow along in the soil. When the south monopoles in the soil meet the north monopoles in roots or a seed, energy is released that causes plant growth. Nitrogen acts as a catalyst to help the process occur.

The more fertile soils are highly paramagnetic and attract more south monopoles (a paramagnetic substance is a non-metallic substance that is weakly attracted by a strong magnet). Clay and volcanic soils are highly paramagnetic and very fertile. Plants grow better in clay flower pots than plastic ones. Many of the ancient stone towers and pyramids around the world are excellent collectors ("antennas") of paramagnetic energy. Plants grow better near them than elsewhere.

Many gardeners and farmers are applying paramagnetic materials to their soils and are getting improved crop germination, growth and seed production. Additives such as compost, certain clays and ground-up rocks of volcanic origin (basalt or granite, for example) usually are paramagnetic. You can buy an electronic meter to measure magnetic energy, or you can do a simple test by filling a plastic tube (or soda straw) with soil or rock powder. Suspend the tube so it can rotate easily (such as hanging it by a thread tied around its center). Then bring a strong magnet near one end of the tube. A paramagnetic material will be attracted to the magnet. A soil that is already fertile and healthy will likely not show much response to addition of paramagnetic materials.

A related idea is "pyramid power," the energy that is apparently focused on the inside and released from the apex and edges of a four-sided pyramid with the same proportions as the pyramid of Cheops in Egypt. Pyramid expert Les Brown (see his book, *The Pyramid*, 1978) states that magnetic energy, cosmic rays and radio waves are focused by a pyramid. He has grown 12- or 13-pound cabbages; 21-inch, four-pound cucumbers; and 50–60 pounds of tomatoes per plant under a pyramid-shaped greenhouse. The same types of plants grown outside produced three-pound cabbages; 14-inch, one-pound cucumbers; and 10–14 pounds of tomatoes per plant. There were no pest problems inside the pyramid. Food placed under a pyramid will gradually dehydrate, but will not rot.

A sound idea. Still another interesting use of natural energy to stimulate plant growth is the use of certain types of sound. Sound waves are a mechanical movement of air molecules or other matter, and are not electromagnetic in nature. Nevertheless, sound transmits energy and can have an effect on living organisms. It is well known that certain types of music (especially rock-and-roll) have harmful effects on our bodies, including our heartbeat, brain waves, and glandular secretions, while most "classical" music has beneficial effects. The wise farmer will tune the radio in his dairy barn to soothing music.

Sound also affects plant growth. A number of early experiments, as well as ancient Hindu lore, have prompted investigations. In 1960, agricultural researcher George E. Smith planted corn and soybeans in two greenhouses. In one he played a recording of "Rhapsody in Blue" continuously. The serenaded plants germinated sooner than those in silence. They also had thicker, greener stems, and ten of the corn plants produced 40 grams of top growth, while the silent-treatment plants only produced 28 grams. The following year a field test of a hybrid corn variety resulted in 137 bushels per acre for musically stimulated corn, compared to 117 for non-stimulated corn.

In the late 1960s, professional musician Dorothy Retallack did carefully controlled experiments (for a college biology course at first) and found that plants will literally wither and die when forced to "listen" to rock-and-roll music (what effect, then, does it have on us?), while classical music, jazz, and sitar music from India stimulated plants. Just a continuous sound tone of 3,000 cycles per second (a low hum) increased growth rates of a variety of plants and caused some house plants to bloom six months early, according to Dr. George Milstein, horticulture teacher at the New York Botanical Garden. Canadian researchers tested early growth (up to eight weeks) of two varieties of wheat at two sound frequencies, 5,000 and 12,000 cycles per second. Plants grew much more vigorously when exposed to the sound, especially one of the varieties (Rideau) at 5,000 c.p.s. The weight of both roots and tops was 250–300% greater than controls, as reported in *Canadian Journal of Botany*, 1968, Vol. 46, p. 1151–1158. They hypothesized that such low levels of energy could cause functional effects in cells if cell parts vibrate at a certain natural frequency. Sound at that frequency would cause much greater vibration (resonance) and add much more energy to the cell.

Dan Carlson, a Minnesota plant breeder, developed a method of combining sound with a special foliar spray to cause fantastic plant growth. Starting with a purple passion house plant which grew 1400 feet long in 2 ½ years (they normally only grow 18 inches tall), Carlson went on to such feats as a 15-foot tomato plant bearing 836 tomatoes; double yield of potatoes; 100–200% yield increase of sweet corn in 30 days, with the corn being much sweeter and better tasting; and 300 or more pods per soybean plant, with 7–10 pods per cluster. Treated seed germinates faster and produces a larger root system. Crops mature earlier; in field tests, potatoes in Minnesota matured by July 1, and grapes matured two weeks early. Cabbages had no insect damage.

Calling his method Sonic Bloom, Carlson uses cassette tapes of his oscillating high frequency sound covered up with Indian sitar or classical music for home or garden use. For agricultural use, he has a sound emitter that attaches to the back of a tractor, which can cover about 35 acres (standing still) with its piercing, canary-like sound. About 15 minutes after the sound is emitted, the crop is sprayed with his special six-ingredient natural growth-stimulating foliar spray. For field crops, three sprayings about two weeks apart are effective. Trees may require more. Tests with radioactive tracers have found that plants absorbed over 700% more spray than controls.

Carlson believes the sound opens up the plants' leaf pores (stomata) so they can "breathe better" and absorb more spray. The sound alone or the spray alone gives good results, but together, they produce much greater results.

The moon's effect. You have no doubt heard of the old custom of planting seed according to the phases of the moon to give better crops.

Modern science has pooh-poohed such "ignorance." But recent research has revealed there is something to it. One researcher, H.S. Burr, found that the moon does indeed affect the physiology of growing seedlings. Much experimenting by M. Thun and H. Heinze in Germany in the 1960s and 1970s discovered that yield is related not to the four quarter phases of the moon, but to the lunar sidereal month of 27.3 days (the movement of the moon in a 360° circle in relation to the positions of fixed stars). They found that seed crops do best if planted when the moon is at 30°, 150°, or 270° apparent longitude. Leaf crops do best if planted when the moon is at 0°, 120°, or 240°. Root crops do best if planted when the moon is at 60°, 180°, or 300°. The longitude of the moon for each day of the year can be found in an annual "Ephemeris" or set of astronomical tables, available in many libraries. However, days when there are eclipses, apogee (when the moon is farthest from the earth), perigee (when the moon is nearest the earth), or other unusual events should be avoided.

We all know the moon affects the tides by gravitational action. It apparently also causes the earth's natural magnetic energy to fluctuate, which may be partly why plants grow better or worse at certain times.

Broadcasting. Another use of weak energies in agriculture (and health) is the controversial practices of radionics and psychotronics. Since each type of atom and molecule radiates a different frequency of energy, it is believed that each plant, animal, or person (or their organs) radiate an energy field that is the sum-total of their individual atoms. This life force is sometimes called "orgone energy." Inanimate materials also radiate at characteristic frequencies. If a plant, animal, or human (or certain cells or organs) is sick, the energy field is altered, and the "vitality" is lowered. A detector machine was developed by Dr. Albert Abrams in the 1920s, which apparently could diagnose human diseases, even from a single drop of blood.

Abrams then reasoned that if sick tissue radiated at a different frequency from healthy, then one should be able to "broadcast" the healthy frequency and help the body to cure itself. Or if there was an infection of germs, broadcasting the frequency of a material that will kill them (such as quinine for malaria) would eradicate the disease. He built such a machine, and it was apparently successful, so much so that the medical establishment rose up to denounce him.

More recent applications of such machines (called radionics machines) have reportedly been able to achieve up to 80 or 90% control of insect pests in fields by broadcasting the frequency of an insecticide, but not actually using any insecticide on the fields (neighboring fields or test plots were decimated by the insects). Even more amazing is that the machine is said to be able to do this at a distance by "treating" an aerial photograph of the field. Radionics operators claim that the emulsion of the photograph captures the frequency of the growing crop. When the USDA investigated such goings on, the USDA teamed with pesticide manufacturers to call radionics a fake, causing the company making the machines to go out of business in the early 1950s. Diseases can supposedly also be diagnosed from a photograph of the patient.

During experimentation in the 1950s, it was found that some people are able to cause plants to grow better without even using a radionics machine—just by mental power, mind over matter. This has developed into what is called psychotronics, the influencing of organisms, with or without the aid of a radionics machine, but guided by the operator's mind.

Because of the possibilities for the use of such methods for evil as well as good (which is admitted by the methods' users), I question the desirability or even the necessity of their use. There are plenty of other effective methods that rely on natural laws and energies alone. In the wrong hands, psychotronics could be used to cause pests and diseases to spread and crops to fail. It could replace germ warfare. It is nothing to play around with.

Alchemy? Can a chicken make calcium from potassium? Can plants and soil bacteria manufacture calcium from silicon, which is part of sand? If such changes in elements seem to defy the laws of chemistry and suggest medieval alchemistry, don't forget that naturally occurring radium disintegrates into lead and helium. Using high-energy atom smashers, nuclear physicists can change one element into another and even create new elements. For example, bombarding nitrogen with helium nuclei (alpha particles) produces oxygen and hydrogen.

Experiments by a number of scientists over the last hundred years indicate that living organisms appear to change or transmutate elements using some mysterious low-level energy that is different from the high-energy that nuclear physicists and atomic bombs use.

Most biologists and biochemists believe that life is essentially chemistry, that living cells are just miniature machines running on chemical energy. Ordinary chemical changes involve the electrons spinning around the central nucleus of an atom. Two atoms of hydrogen, for example, combine with one atom of oxygen to form a molecule of water by sharing electrons.

In other cases, one or more electrons are transferred from one atom to another (forming ions).

But although scientists can "take apart" a cell and find out what elements it is made of, they cannot create life. Life seems to contain an extra "spark" of energy—it is more than the sum of its parts.

Beginning in the 1930s, French scientist Louis Kervran did some amazing experiments. He fed a chicken only oats, then measured the amount of calcium in its eggs and droppings. Kervran found that the chicken produced four times as much calcium as was in the oats. A biochemist might suggest that the bird took the calcium from its skeleton. A chicken deprived of calcium in its food will lay soft-shelled eggs after several days; if the chicken is fed potassium, it will lay normal calcium-shelled eggs. Oats are rich in potassium. So how can a newly-hatched chick's skeleton contain four times more calcium than was in the egg *and* have the shell's calcium content remain the same? These observations, plus many other experiments by Kervran and others on such diverse organisms as microbes, plants, animals, and humans, have shown that living organisms can apparently change one element into another, sometimes called biological transmutation.

They may do this by changes of the atomic nuclei, rather than the outer electrons. The nuclei of elements other than hydrogen contain both neutrons and protons. The number of protons is called the atomic number of that element, which is written as a small number to the lower left of the element's symbol. For example, potassium has an atomic number of 19 (written $_{19}K$), hydrogen is one ($_1H$), and calcium is twenty ($_{20}Ca$). If you take a potassium atom and add a hydrogen atom, you get a calcium atom ($_{19}K + _1H = _{20}Ca$). Calcium can also be formed by adding oxygen to magnesium ($_{12}Mg + _8O = _{20}Ca$), or by adding carbon to silicon ($_{14}Si + _6C = _{20}Ca$). Or you can remove hydrogen from calcium to get potassium, or combine sodium and oxygen ($_{11}Na + _8O = _{19}K$). This may be partly how soil microbes and some plants can enrich poor soil with certain nutrients, and why rotations and companion crops work so well. Kervran believed microorganisms are of utmost importance in correcting soil imbalances, while overuse of synthetic N-P-K fertilizers leads to imbalances and unhealthy crops.

So, while you may have trouble swallowing much of the above, it is fun to speculate. There is always more to learn!

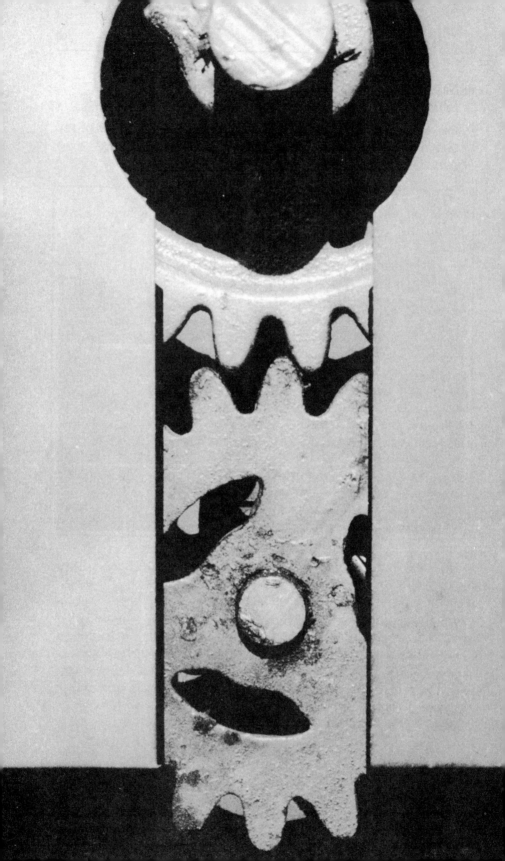

Chapter 21

A COG IN THE WHEEL?

I f you have read this far, you are obviously not an average person—you care. You are concerned about problems and want to find solutions. You are not just another cog in the wheel of modern agriculture. You want to think for yourself.

We have taken a journey searching for seldom-known aspects of agriculture today. We have taken many paths and detours. We have looked at nature's system and man's system. We have seen how a plant makes food and how a manufacturer makes a fertilizer formula. We have found some interesting and surprising things.

I hope you have become convinced that for the most part, "modern" agriculture is misguided, heading in the wrong direction. There is a better way! I hope you are convinced that "natural-system" agriculture is the way to go. It is the wave of the future. It is ahead of its time. It is actually more scientific than what the "experts" call scientific agriculture, for it is based on natural laws and systems, and seeks to work with them, not against them.

Natural-system farming is not impractical—it works! Once you get your soil turned around and into a healthy state, things will really happen: high yields of crops with very high nutritional value—quality—and healthy, high-producing animals. And your family's health will also improve. And probably at no more expense than what you spend now for fertilizers, herbicides and pesticides, and vet and doctor bills. Of course there will be a transition period while things are turning around, while the poisons are being cleaned out of your soil and the beneficial organisms are being reestablished. That is a delicate time when you need patience and optimism, and plenty of reliable advice. But you will see encouraging results right from the beginning, and crops will improve steadily, year after year.

It takes guts. It takes courage to break out of the mold. To be different from your neighbors. To try something new. To change.

Agriculture has to change. The present system is leading us downward in an ever-increasing spiral. It takes concerned and far-sighted individuals to break out of the system and go on to better things. You can do it—if you really want to!

Glossary

aerobic—with oxygen present (see anaerobic)

anaerobic—without oxygen (see aerobic)

anion—a negatively-charged ion (see cation)

atom—the smallest particle of matter (an element) which has the properties of that element

cation—a positively-charged ion (see anion)

cell—the smallest unit of living organisms

chelate—a water soluble organic compound that can hold a metal cation

chemical—any particular kind of matter (see atom, compound, element, ion, molecule)

colloid, colloidal—a material of very small particle size (less than 0.002 mm, or 1/5000 inch), but larger than most molecules; has a large surface area

compound—a material composed of two or more elements

ecology—the study of organisms and their environment, and the interrelationships among them

ecosystem—a total ecological system composed of all of the living organisms plus the non-living environment

element—the simplest type of matter, composed of only one type of atom

enzyme—a substance which increases the speed of metabolic or other chemical changes, usually inside living cells

foliar—referring to leaves

gene—the smallest unit of heredity; found on chromosomes in every cell; determines the inherited characteristics of living organisms

hormone—a substance secreted by a plant or animal that regulates developmental processes: metabolism, growth, and reproduction

humus—a colloidal material resulting from decomposition of plant or animal organic matter

inorganic—*chemical definition*: chemical compound not containing carbon; *ordinary definition*: not produced by living organisms (see organic)

ion—an electrically charged atom or group of atoms

metabolism, metabolic—any of the chemical activities of living cells or organisms

microbe—same as microorganism

microorganism—a microscopic organism

molecule—the smallest particle of a compound, composed of two or more atoms or elements

organic—*chemical definition*: chemical compound containing carbon; *ordinary definition*: referring to or produced by living organisms (see inorganic)

organism—any living thing, whether microbe, plant, or animal

osmosis—seeping (diffusion) of water through a thin membrane, from a higher concentration to a lower concentration of water

parasite—an organism that lives on or in the body of another organism and derives its nutrition from the other organism

pathogen—a disease-causing microorganism

pH—a measure of acidity or alkalinity; with pH 7 being neutral, pH 0 to 7 acid, and pH 7 to 14 alkaline

photosynthesis—producing food (sugar) by green plants, using sunlight

physiology, physiological—the functioning of cells or organisms

predator—an organism (animal) that gets its food by killing and eating other animals

respiration—releasing energy from food by breaking it down inside the organism's cells

salt—a chemical compound formed by combining ions from an acid and a base

species—a certain kind of organism

stoma, stomata, stomates—tiny openings on the surface of leaves that allow gas exchange

tilth—a crumbly soil texture

tissue—a group of cells having a particular function

translocation—transfer of food or other products from one part of a plant to another

BASIC CHEMISTRY

Although we cannot present an entire chemistry course here, we will give enough of the principles and terminology of chemistry to aid in understanding soils, fertilizers, and plant functions. First some terms and concepts.

1. *Chemical.* Not a "bad word," but merely the name for any kind of matter, whether gas, liquid, or solid. All matter, including yourself, is composed of chemicals.

2. *Organic vs. inorganic.* "Organic" is sometimes used by farmers, gardeners, and health enthusiasts as synonymous with "natural," or produced by living organisms (this was the original meaning). The chemical definition of organic is carbon-containing compounds; all other compounds are called inorganic (most inorganic chemicals are "natural," such as oxygen, nitrogen, and sulfur).

3. *Compound.* A pure substance that can be decomposed into two or more different elements. Examples of compounds include water, carbon dioxide, and sugar.

4. *Element.* A type of matter that cannot be decomposed into simpler substances. There are only 90 elements that occur in nature (plus others made by man). Each element is given a name and an abbreviation, or chemical symbol, such as H for hydrogen, O for oxygen, etc. (see box).

5. *Atom.* Smallest particle of an element that can enter into chemical changes (but composed of sub-particles: protons, neutrons, and electrons, forming a "miniature solar system").

6. *Molecule.* Smallest particle of a compound, or a stable combination of atoms of one element, such as O_2, Cl_2, H_2O, CO_2, H_2CO_3.

7. *Ion.* An atom or group of atoms carrying an electrical charge. There are two kinds, cations (positively charged, such as H^+, Na^+, Ca^{++}) and anions (negatively charged, such as Cl^-, O^{--}, SO_4^{--}). The number of + or − charges determines how many ions are needed to combine to form a molecule, such as H_2O, $CaCl_2$, CaO. The + and − charges must equal zero.

8. *Bonds.* Atoms and ions are held together by various kinds of chemical bonds, which are an attractive force, a type of stored energy, which is released when the bonds are broken.

Reactions and equations. Many chemical changes (reactions) involve compounds that act as acids and bases. Typical acids contain hydrogen ions (H^-); typical bases contain hydroxyl ions (OH^-). Acids react with bases to form salts and water. We can write the reaction using chemical formulas; the reaction is called an equation, and the number of atoms of each element must be balanced on each side of the arrow. For example:

hydrochloric acid + sodium hydroxide yields sodium chloride + water
$$HCl + NaOH \rightarrow NaCl + H_2O$$

Another reaction, involving sulfuric acid is:
sulfuric acid + sodium hydroxide yields sodium sulfate + water
$$H_2SO_4 + 2NaOH \rightarrow Na_2SO_4 + 2H_2O$$

Other types of chemical reactions are similarly written:

1. Two gases combine: hydrogen + oxygen yields water
$$2H_2 + O_2 \rightarrow 2H_2O$$

2. Decomposition: ammonium chloride yields (when heated) ammonia + hydrogen chloride
$$NH_4Cl \rightarrow NH_3 + HCl$$

3. Simple replacement: iron + sulfuric acid yields ferrous sulfate + hydrogen
$$Fe + H_2SO_4 \rightarrow FeSO_4 + H_2$$

4. Double decomposition: calcium chloride + sulfuric acid yields calcium sulfate + hydrogen chloride
$$CaCl_2 + H_2SO_4 \rightarrow CaSO_4 + 2HCl$$

Chemical Elements
Essential to Plants

element	chemical symbol	forms available to plants
carbon	C	CO_2 (carbon dioxide)
oxygen	O	O_2 (oxygen gas), H_2O (water)
hydrogen	H	H_2O (water)
nitrogen	N	NO_3^- (nitrate), NH_4^+ (ammonium)
calcium	Ca	Ca^{++} (calcium ion)
phosphorus	P	$H_2PO_4^-$ (dihydrogen phosphate, orthophosphate), HPO_4^{--} (hydrogen phosphate)
magnesium	Mg	Mg^{++} (magnesium ion)
potassium	K	K^+ (potassium ion)
sulfur	S	SO_4^{--} (sulfate)
chlorine	Cl	Cl^- (chloride)
iron	Fe	Fe^{+++} (ferric), Fe^{++} (ferrous)
copper	Cu	Cu^{++} (cupric), Cu^+ (cuprous)
zinc	Zn	Zn^{++} (zinc ion)
boron	B	BO_3^{--} (borate), $B_4O_7^{--}$ (perborate)
manganese	Mn	Mn^{++} (manganous)
molybdenum	Mo	MoO_4^{--} (molybdate)

RESOURCES & ORGANIZATIONS

The following list of sources of information and/or contact people is by no means complete. Many others, especially local ones and others in countries outside the United States can be found, for instance, on the Internet.

Organizations

1. **Acres U.S.A.** A national magazine promoting sustainable (ecological) agriculture. Holds an annual educational conference, and sells a large variety of relevant books. P.O. Box 91299, Austin, TX 78709-1299; 5321 Industrial Oaks Blvd., Ste. 128, Austin, TX 78735. Phone: (800) 355-5313; (512) 892-4400. Web: *www.acresusa.com.*

2. **American Farmers Trust (AFT).** Seeks to stop the loss of productive farmland and promotes environmentally healthy farming practices. 1200 18th St. NW, Ste. 800, Washington, D.C. 20036. Phone: (800) 431-1499; (202) 331-7300. Web: *www.farmland.org.*

3. **Biodynamic Farming and Gardening Association.** Promotes the biodynamic method of agriculture. Building 1002B, Thoreau Center, The Presidio, P.O. Box 29135, San Francisco, CA 94129. Phone: (888) 516-7797. Web: *www.biodynamics.com.*

4. **California Sustainable Agriculture Working Group.** A network of organizations promoting sustainable agriculture and food systems. P.O. Box 1599, Santa Cruz, CA 95061. Phone: (831) 457-2815. Web: *www.calsawg.org.*

5. **Center for Agroecology and Sustainable Food Systems (CASFS).** Seeks to increase ecological sustainability and social justice in food and agricultural systems. University of California-Santa Cruz, 1156 High St., Santa Cruz, CA 95064. Phone: (831) 459-3240. Web: *http//socialsciences.ucsc.edu/casfs/.*

6. **Center for Food Safety (CFS).** An advocacy organization seeking to curtail the industrial agriculture system and promote sustainable alternatives. 660 Pennsylvania Ave. SE, Ste. 302, Washington, D.C. 20003. Phone: (202) 547-9359. Web: *www.centerforfoodsafety.org; www.foodsafetynow.org.*

7. **Ecological Farming Association (EFA).** Promotes ecologically sound agriculture. 406 Main St., Ste. 313, Watsonville, CA 95076. Phone: (408) 763-2111. Web: *www.csa-efc.org.*

8. **Food Alliance.** Promotes sustainable agricultural practices. 1829 Alberta, Ste. 5, Portland, OR 97211. Phone: (503) 493-1066. Web: *www.foodalliance.org.*

9. **Friends of the Earth.** Seeks to preserve the health and environmental diversity of the planet for future generations. 1025 Vermont Ave. NW, Ste. 300, Washington, D.C. 20005. Phone: (800) 843-8687; (202) 783-7400. Web: *www.foe.org.*

10. **Institute for Agriculture and Trade Policy.** Promotes family farms, rural communities and ecosystems. 2105 First Ave. S, Minneapolis, MN 55404. Phone: (612) 870-0453. Web: *www.iatp.org.*

11. **International Federation of Organic Agriculture Movements (IFOAM).** Promotes worldwide adoption of ecologically, socially and economically sustainable systems based on the principles of organic agriculture. Charles-de-Gaulle-Str. 5, 53113 Bonn, Germany. Phone: +49-228-92650-10. Web: *www.ifoam.org.*

12. **Kerr Center for Sustainable Agriculture.** Promotes a more sustainable agriculture for farmers and ranchers in Oklahoma and region. P.O. Box 588, Poteau, OK 74953. Phone: (918) 647-9123. Web: *www. kerrcenter.com.*

13. **The Land Institute.** Seeks to develop an agricultural system with the ecological stability of the prairie and grain yields equivalent to annual crops. 2440 E. Water Well Rd., Salina, KS 67401. Phone: (785) 823-5376. Web: *www.landinstitute.org.*

14. **Land Stewardship Project.** Promotes sustainable agriculture, sustainable communities and regional food systems. 2200 4th St., White Bear Lake, MN 55110. Phone: (651) 653-0618. Web: www. *landstewardshipproject.org.*

15. **Leopold Center for Sustainable Agriculture.** Seeks to reduce the negative impacts of agriculture on natural resources and rural communities and to develop profitable farming systems that conserve natural resources. 209 Curtiss Hall, Iowa State University, Ames, IA 50011. Phone: (515) 294-3711. Web: *www.leopold.iastate.edu.*

16. **Michael Fields Agricultural Institute.** Seeks to develop a sustainable agriculture through research, education, technical assistance and public policy. W2493 County Rd. ES, P.O. Box 990, East Troy, WI 53120. Phone: (262) 642-3303. Web: *www.michaelfieldsaginst.org.*

17. **Midwest Sustainable Agriculture Working Group.** A network of organizations promoting sustainable agricultural and food systems in the Midwest. 110 Maryland Ave. NE, Ste. 209, Washington, D.C. 20002. Phone: (202) 547-5754. Web: *msawg.org.*

18. **National Family Farm Coalition.** A national center for organizations seeking to preserve and strengthen family farms and oppose corporate agriculture. 110 Maryland Ave. NE, Ste. 307, Washington, D.C. 20002. Phone: (202) 543-5675. Web: *vww.nffc.net.*

19. **National Farmers Organization (NFO).** Seeks to improve agricultural commodity prices by serving as a bargaining agent for

members. 528 Billy Sunday Rd., Ste. 100, Ames, IA 50010. Phone: (800) 247-2110. Web: *www.nfo.org.*

20. **National Organization for Raw Materials (NORM).** Seeks to study and inform about raw material economics and return the U.S. economy to widespread and stable prosperity. 680 E. 5 Point Hwy., Charlotte, MI 48813. Phone: (517) 543-0111. Web: *www.normeconomics.com.*

21. **National Sustainable Agricultural information (ATTRA).** Provides information and technical assistance. P.O. Box 3657, Fayetteville, AR 72702. Phone: (800) 346-9140. Web: *www.attra.org.*

22. **Northeast Sustainable Agriculture Working Group.** A network of organizations promoting sustainable agricultural and food systems in the Northeast. P.O. Box 11, Belchertown, MA 01007. Phone: (413) 323-9878. Web: *www.nesawg.org.*

23. **Organic Consumers Association (OCA).** An online and grassroots public interest organization promoting health, justice and sustainability. 6771 South Silver Hill Dr., Finland, MN 55603. Phone: (218) 226-4164. Web: *www.organicconsumers.org.*

24. **Organic Farming Research Foundation.** Sponsors research on organic farming practices and educates the public about organic farming issues. P.O. Box 440, Santa Cruz, CA 95061. Phone: (831) 426-6606. Web: *www.ofrf.org.*

25. **Organic Farmers Marketing Association.** Assists farmers in promoting and marketing organic products, including a directory of certifiers. P.O. Box 2407, Fairfield, IA 52556. Phone: (515) 472-3272. Web: *www.iquest.net/ofma.*

26. **Practical Farmers of Iowa (PFI).** Promotes and does research on ecologically sound and profitable agriculture. P.O. Box 349, Ames, IA 50010. Phone: (515) 232-5661. Web: *www.practicalfarmers.org.*

27. **Rodale Institute.** Promotes a worldwide regenerative food system that renews environmental and human health. 611 Seigfriedale Rd., Kutztown, PA 19530. Phone: (610) 683-1400. Web: www. *rodaleinstitute.org.*

28. **Seed Savers Exchange.** Promotes preserving and exchanging genetically diverse old, traditional food crops. 3076 N. Winn Rd., Decorah, IA 52101. Web: *www.seedsavers.org.*

29. **Southern Sustainable Agriculture Working Group.** A network of organizations promoting sustainable agriculture in the South. P.O. Box 324, Elkins, AR 72727. Phone: (479) 587-0888. Web: *www.ssawg.org.*

30. **Sustainable Agriculture Research and Education Program, USDA (SARE).** Promotes farming systems that are profitable, environmentally sound and good for communities through a nationwide research and education grants program; part of the USDA's Cooperative State Research, Education and Extension Service. USDA-CSREES, Stop 2223, 1400 Independence Ave. SW, Washington, D.C. 20250-2223. Phone: (202) 720-6071. Web: *www.sare.org.*

31. **Western Sustainable Agriculture Working Group.** A network of organizations promoting sustainable agricultural and food systems in the West. P.O. Box 59, Victor, MT 59702. Phone: (406) 642-3601. Web: *www.westernsawg.org.*

32. **Wild Farm Alliance.** Promotes ecological farming that protects and restores the natural environment through sustainable agriculture and biodiversity conservation. P.O. Box 2570, Watsonville, CA 95077. Phone: (831) 761-8408. Web: *www.wildfarmalliance.org.*

33. **World Sustainable Agriculture Association.** Disseminates scientific and technical information on sustainable agriculture and marketing. 8554 Melrose Ave., West Hollywood, CA 90060. Phone: (310) 657-7202. E-mail: *pmaddeni@aol.com.*

Directories

1. **Local Harvest,** a directory of local farmers' markets and community-supported agriculture (CSA) organizations. Web: *www.localharvest.com.*

2. **Organic Farming Research Foundation,** a directory of organic certifiers. Web: *http://ofrf.org/publications/certifier.html.*

3. **Organic Trade Association,** a business association that aims to promote and protect the growth of organic trade to benefit the environment, farmers, the public and the economy. Web: *www.ota.com.*

Index

0-0-60, 174, 176
0-0-62, 174, 176
2,4,5-T (Agent Orange), 239
2,4-D, 230, 233
acid soils, 277
acidity, 181
acidity, neutralizing, 184
acids, 39, 351
actinomycetes, 45, 97
activated water, 329
active acidity, 181, 182
active fraction, 190
aerobic, 51, 347
aerobic decomposition, 100
aflatoxin, 296-297
Agent Orange, 239
aggregates, 38
agricultural institutions, 5
agriculture, failings, 124
agriculture, modern, 2
agriculture, unnatural, 325
air, 38
Albrecht, Dr. William, 295
aldicarb (TEMIK), 231
algae, 274
alien invaders, 29-30
alkaline, 39
alkaline soils, 277-278
allelopathy, 254, 257, 330
amino acids, 91-92, 243
ammonia, 162, 170-171
ammonification, 98
ammonium, 150, 157

ammonium hydroxide (NH$_4$OH),
   170
anabolism, 86
anaerobic, 51, 53-54, 347
anaerobic conditions, chart, 55
anaerobic decomposition, 98
anaerobic soil, 52
anaerobic soil, causes, 55-56
anaerobic soil, corrections, 56-57
anhydrous ammonia, 167,
   170-171
animal health, 253
animal matter, 191
anions, 320, 347, 351
anthroposophy, 319
Archer, S.G., 11
Aspergillus flavus, 296
Aspergillus parasiticus, 296
atom, 347, 350
atomic nuclei, 343
ATP (adenosine triphosphate), 72
atrazine, 232

bacteria, 43, 45, 97
bacteria and fungi ratio, 96
bacteria, cryophilic, 100
bacteria, denitrifying, 49
bacteria, mesophilic, 100-101
bacteria, thermophilic, 101
bacterial inoculants, 274
balance-sheet theory, 3, 141-142
base exchange, 62, 182
base exchange capacity, 142

base saturation, 142, 186
bases, 351
benzene hexachloride (BHC), 230
Best Land Use, chart, 324
biodiesel, 316
biodynamic agriculture, 318-319
biofuels, 316-317
biological products, 286
biological transmutation, 343
biological value (BV), 90
biomass, 316-317
biotic pyramid, 81
Blake, J.L., 197
bonds, 351
botanicals, 236, 252
breast cancer, 239
Brix, 292, 293
Brix, chart, 294
Brummer, E. Charles, 124
Bt, 299

CaCO₃, 277
CaCO₃•MgCO₃, 277
calcium, 157, 186, 307
calcium carbonate, 277
calcium lime, 275
calcium magnesium carbonate, 277
calcium sulfate, 278
calcium-potassium ratio, 306-307
Callahan, Dr. Philip, 245, 338
calorie, 89
carbon dioxide, 52, 73-74
carbonaceous, 157, 306-307
carbon-to-nitrogen ratio, 101, 191, 194
Carlson, Dan, 340
castings, 99
catabolism, 86

catalyst water, 329
catalysts, 72
cation exchange, 62, 182
cation exchange capacity (CEC), 142
cations, 142, 181-182, 320, 351
cell, 347
cellular respiration, 74
chelates, 62, 66-67, 347
chelates, synthetic, 67
chelators, 34
chemical, 347, 350
chemical energy, 71
chemical reactions, 351
chemical, side effects, 14, 17-18
chemicals, usage, 13
chlorine, 162, 174, 176
chlorophyll, 71, 104
climate change, 30-31
cobalt, 303
Collapse: How Societies Choose to Fail or Succeed, 118
colloidal phosphate, 286
colloids, 38, 347
compacted soil, 285
compost, 56, 190-194, 278, 319
compost, application, 194
compost, ditch irrigation, 194
compost, microorganisms, 192-193
compost, moisture, 192
compost tea, 96, 332
compost, temperature, 192
composting, 103-104
compound, 347, 350
continuous cropping, 202-203, 280-281
corn, dent, 294
corn stalk rot, 235, 259-263

corn, standard weight, 291
corn, yields, 290
cosmic pipe, 336
cover crops, 273
cow, feed pounds, 291-292
crop losses, 14
crop rotations, 202, 203, 280-281
crop, variety, 19-20
crude protein (CP), 90-91, 214, 290-291
cryophilic bacteria, 100
cultivation, 330

DDT ppm, diagram, 238
dead zones, 30
decomposers, 95
deforestation, 29
denitrification, 166
desertification, 9
detoxification, 105
Diamond, Jared, 118
diatomaceous earth, 252
diethylstilbestrol (DES), 15
digestible energy (DE), 89
digestion, 83-84
dioxin, 239
Diplodia maydis, 259
disease, 82
disking, 207
DNA, 173
dolomite, 184, 186
dolomitic lime, 275
dry soils, 278-279

earthworms, 96-97, 99-100, 274, 283
ecology, 25, 347
ecosystems, 26, 325, 347

EDTA (ethylene diamine tetraacetic acid), 67
electrical charge, 320
electromagnetic fields, 334
electromagnetic radiation, 332
electromagnetic spectrum, diagram, 333
electromagnetic vibrations, 335
element, 347, 350
eloptic energy, 336
endocrine disrupters, 16-17
energy carriers, 72
energy, chemical, 71
energy production, 87
energy storage, 90
enzymes, 72, 348
Ephemeris, 341
equation, biological value, 90
equation, digestion coefficient, 89
equation, K_2O to K, 166
equation, metabolizable energy, 89
equation, net energy, 90
equation, P_2O_5 to P, 166
erosion, 8-9, 37
essential amino acids, 90
essential minerals, 88
ethanol, 316-317
ethylene gas, 247
eutrophication, 18
excretion, 86
exudate, 64-65

farm dollar ratio, 114
farm economy, 111-112
farm facts, 107-108
Farmer's Every-day Book, The, 197
fascism, 121-122
Federalist Society, 123
fermentation, 54-55, 98

pharming, 22
phloem, 68, 76
phosdrin, 240
phosphate, 179
phosphoric acid, 160, 166
Phosphorus Sources, table, 175
phosphorus, 72, 166, 172, 173-174, 307
photosynthesis, 69, 71, 149, 348
physiology, 348
phytoalexins, 243
phytotoxins, 257
Pimentel, Dr. David, 14, 17
plant barriers, 242
plant elements, 352
plant matter, 191
plastics, 16
plowing, 208-209
poison rain, 231
positive ions, 333
potash, 166, 174, 176
potassium, 157, 166, 172, 174
potassium chloride, 174, 176
potential acidity, 182
prairie soils, 278-279
predator, 349
pre-humus, 190
"Project for the New American Century," 123
protein, 307
protein content, 289
proteinaceous, 157, 306-307
psychotronics, 342
pyramid power, 339

rabbits, 214
radionics, 341-342
rapid growth, 150

Reams Theory of Biological Ionization (RBTI), 320
Reams, Dr. Carey, 320
reduced tillage, 205, 206
refractometer, 292-293
repellent chemicals, 242-243
respiration, 349
rGBH, 17
rhizosphere, 64
Rhyzobium, 43, 45
ridge planting, 207-208
ridge-tilling, 283
RNA, 173
rock fertilizers, 272
rock phosphate (0-2-0), 179, 285-286
roots, 64
roots, rye plant, 61
row crops, 203
ruminants, digestion, 84-85

saline soils, 277-278
salt fertilizer, 61, 62, 96, 152, 163
salt fertilizer, problems, 198
salt index, 156
salts, 349, 351
sandy soils, 172
sea level, 31
seaweed, 160, 274, 287
Second American Agricultural Revolution, 2
secondary metabolites, 243
seed treatments, 328
sickness, 82
slogans, 121
socialism, 120
soft rock phosphate, 286
soil, 37-38
soil air, 51

Also from Acres U.S.A.

Reproduction & Animal Health

BY CHARLES WALTERS & GEARLD FRY

This book represents the combined experience and wisdom of two leaders in sustainable cattle production. Gearld Fry offers a lifetime of practical experience seasoned by study and observation. Charles Walters draws on his own observations as well as interviews with thousands of eco-farmers and consultants over the past four decades. The result is an insightful book that is practical in the extreme, yet eminently readable. In this book you will learn: how to "read" an animal, what linear measurement is, why linear measurement selects ideal breeding stock, the nuances of bull fertility, the strengths of classic cattle breeds, the role of pastures, the mineral diet's role in health. *Softcover, 222 pages. ISBN 0-911311-76-9*

Hands-On Agronomy

BY NEAL KINSEY & CHARLES WALTERS

The soil is more than just a substrate that anchors crops in place. An ecologically balanced soil system is essential for maintaining healthy crops. This is a comprehensive manual on soil management. The "whats and whys" of micronutrients, earthworms, soil drainage, tilth, soil structure and organic matter are explained in detail. Kinsey shows us how working with the soil produces healthier crops with a higher yield. True hands-on advice that consultants charge thousands for every day. Revised, third edition. *Softcover, 352 pages. ISBN 0-911311-59-9*

Hands-On Agronomy Video Workshop

Video Workshop

BY NEAL KINSEY

Neal Kinsey teaches a sophisticated, easy-to-live-with system of fertility management that focuses on balance, not merely quantity of fertility elements. It works in a variety of soils and crops, both conventional and organic. In sharp contrast to the current methods only using N-P-K and pH and viewing soil only as a physical support media for plants, the basis of all his teachings are to feed the soil, and let the soil feed the plant. The Albrecht system of soils is covered, along with how to properly test your soil and interpret the results. *80 minutes.*

To order call 1-800-355-5313 or order online at www.acresusa.com

Agriculture in Transition

BY DONALD L. SCHRIEFER

Now you can tap the source of many of agriculture's most popular progressive farming tools. Ideas now commonplace in the industry, such as "crop and soil weatherproofing," the "row support system," and the "tillage commandments," exemplify the practicality of the soil/root maintenance program that serves as the foundation for Schriefer's highly-successful "systems approach" farming. A veteran teacher, lecturer and writer, Schriefer's ideas are clear, straightforward, and practical. *Softcover, 238 pages. ISBN 0-911311-61-0*

From the Soil Up

BY DONALD L. SCHRIEFER

The farmer's role is to conduct the symphony of plants and soil. In this book, learn how to coax the most out of your plants by providing the best soil and re-moving all yield-limiting factors. Schriefer is best known for his "systems" approach to tillage and soil fertility, which is detailed here. Managing soil aeration, water, and residue decay are covered, as well as ridge planting systems, guidelines for culti-vating row crops, and managing soil fertility. Develop your own soil fertility system for long-term productivity. *Softcover, 274 pages. ISBN 0-911311-63-7*

Science in Agriculture

BY ARDEN B. ANDERSEN, PH.D., D.O.

By ignoring the truth, ag-chemical enthusiasts are able to claim that pesticides and herbicides are necessary to feed the world. But science points out that low-to-mediocre crop production, weed, disease, and insect pressures are all symptoms of nutritional imbalances and inadequacies in the soil. The progressive farmer who knows this can grow bountiful, disease- and pest-free commodities without the use of toxic chemicals. A concise recap of the main schools of thought that make up eco-agriculture — all clearly explained. Both farmer and professional consultant will benefit from this important work. *Softcover, 376 pages. ISBN 0-911311-35-1*

Bread from Stones

BY JULIUS HENSEL

This book was the first work to attack Von Liebig's salt fertilizer thesis, and it stands as valid today as when first written over 100 years ago. Conventional agri-culture is still operating under misconceptions disproved so eloquently by Hensel so long ago. In addition to the classic text, comments by John Hamaker and Phil Callahan add meaning to the body of the book. Many who stand on the shoul-ders of this giant have yet to acknowledge Hensel. A true classic of agriculture. *Softcover, 102 pages. ISBN 0-911311-30-0*

To order call 1-800-355-5313 or order online at www.acresusa.com

The Secret Life of Compost

BY MALCOLM BECK, WITH COMMENTARY BY CHARLES WALTERS

 We don't need to poison the earth in order to grow better food, and what is harmful to the environment when improperly disposed of often can be turned back to the soil in a beneficial way through composting — if you know how. Here's how. Malcolm Beck's Garden-Ville is one of the largest commercial composting operations in the country. He shares his insight into the processes of decay that can transform everything from lawn trimmings to sewer sludge into life-giving earth. Coupled with Beck's insight into nature and practical advice are remarks from Charles Walters, author and founder of *Acres U.S.A. Softcover, 150 pages. ISBN 0-911311-52-1*

Fletcher Sims' Compost

BY CHARLES WALTERS

 Covers the optimal conditions for converting plant and animal wastes into compost by balancing the correct ratio of raw materials, using the correct microorganisms and moisture content, proper pile or windrow construction, and efficient mixing. Fletcher Sims, the Dean of Composters, has elevated the "art" of good composting to a "science." Explains not only the complexities of commercial-scale compost production, but also the benefits of the use of this gentle fertilizer. A book that really draws you in, it is a combination of a biography and technical guide written by the founder of *Acres U.S.A. Softcover, 247 pages. ISBN 0-911311-43-2*

A Farmer's Guide to the Bottom Line

BY CHARLES WALTERS

 This book is the culmination of Walters' lifetime of experience, written in his honest, straight-ahead style, outlining how the small farmer-entrepreneur can find his way to a profitable bottom line. The book provides how-to information on each step from planning to implementation of business practices for the eco-friendly farm and includes examples of people who are making a living, and a profit, by demanding a fair price for their labor. Whether you are considering taking up farming as an occupation or just interested in the economics and history of farming, this book is a must-read. *Softcover, 212 pages. ISBN 0-911311-71-8*

How to Grow World Record Tomatoes

BY CHARLES H. WILBER

 For most of his 80+ years, Charles Wilber has been learning how to work with nature. In this almost unbelievable book he tells his personal story and his philosophy and approach to gardening. Finally, this Guinness world record holder reveals for the first time how he grows record-breaking tomatoes and produce of every variety. Detailed step-by-step instructions teach you how to grow incredible tomatoes — and get award-winning results with all your garden, orchard, and field crops! Low-labor, organic, bio-intensive gardening at its best. *Softcover, 132 pages. ISBN 0-911311-57-2*

Eco-Farm: An Acres U.S.A. Primer

BY CHARLES WALTERS

 In this book, eco-agriculture is explained — from the tiniest molecular building blocks to managing the soil — in terminology that not only makes the subject easy to learn, but vibrantly alive. Sections on NP&K, cation exchange capacity, composting, Brix, soil life, and more! *Eco-Farm* truly delivers a complete education in soils, crops, and weed and insect control. This should be the first book read by everyone beginning in eco-agriculture . . . and the most shop-worn book on the shelf of the most experienced. *Softcover, 476 pages.* ISBN 0-911311-74-2

Weeds: Control Without Poisons

BY CHARLES WALTERS

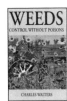 For a thorough understanding of the conditions that produce certain weeds, you simply can't find a better source than this one — certainly not one as entertaining, as full of anecdotes and homespun common sense. It contains a lifetime of collected wisdom that teaches us how to understand and thereby control the growth of countless weed species, as well as why there is an absolute necessity for a more holistic, eco-centered perspective in agriculture today. Contains specifics on a hundred weeds, why they grow, what soil conditions spur them on or stop them, what they say about your soil, and how to control them without the obscene presence of poisons, all cross-referenced by scientific and various common names, and a new pictorial glossary. *Softcover, 352 pages. ISBN 0-911311-58-0*

The Biological Farmer
A Complete Guide to the Sustainable
& Profitable Biological System of Farming
BY GARY F. ZIMMER

 Biological farmers work with nature, feeding soil life, balancing soil minerals, and tilling soils with a purpose. The methods they apply involve a unique system of beliefs, observations and guidelines that result in increased production and profit. This practical how-to guide elucidates their methods and will help you make farming fun and profitable. *The Biological Farmer* is the farming consultant's bible. It schools the interested grower in methods of maintaining a balanced, healthy soil that promises greater productivity at lower costs, and it covers some of the pitfalls of conventional farming practices. Zimmer knows how to make responsible farming work. His extensive knowledge of biological farming and consulting experience come through in this complete, practical guide to making farming fun and profitable. *Softcover, 352 pages. ISBN 0-911311-62-9*

To order call 1-800-355-5313 or order online at www.acresusa.com

Alternative Treatments for Ruminant Animals

BY PAUL DETTLOFF, D.V.M.

Drawing on 36 years of veterinary practice, Dr. Paul Dettloff presents an natural, sustainable approach to ruminant health. Copiously illustrated chapters "break down" the animal into its interrelated biological systems: digestive, reproductive, respiratory, circulatory, musculoskeletal and more. Also includes a chapter on nosodes, with vaccination programs for dairy cattle, sheep and goats. An information-packed manual from a renowned vet and educator. *Softcover, 260 pages. ISBN 0-911311-77-7*

Grass, the Forgiveness of Nature

Exploring the miracle of grass, pastures & grassland farming

BY CHARLES WALTERS

What is the most important plant in the world? In terms of nutritive content, function within the ecosystem, and even medicinal properties, the answer to this question may very well be *grass*. In this wide-ranging survey of grass forages and pastureland, Charles Walters makes the case that grass is not just for cows and horses — that in fact it is the most nutritious food produced by nature, as well as the ultimate soil conditioner. You will learn from traditional graziers who draw on centuries of wisdom to create beautiful, lush, sustainable pastures, as well as cutting-edge innovators who are using such methods as biodynamics and sea-solids fertilization to create some of the healthiest grasslands in the world. Leading agronomists not only explain the importance of grasses in our environment, they also share practical knowledge such as when to look for peak levels of nutrition within the growing cycle and how to use grass to restore soil to optimum health. A must-read for anyone interested in sustainable, bio-correct agriculture, this information-packed volume is a comprehensive look at an essential family of plants. *Softcover, 320 pages. ISBN 0-911311-89-0*

Soil, Grass & Cancer

BY ANDRÉ VOISIN

Almost a half-century ago, André Voisin had already grasped the importance of the subterranean world. He mapped the elements of the soil and their effects on plants, and ultimately, animal and human life as well. He saw the hidden danger in the gross oversimplification of fertilization practices that use harsh chemicals and ignore the delicate balance of trace minerals and nutrients in the soil. With a volume of meticulously researched information, Voisin issues a call to agricultural scientists, veterinarians, dietitians and intelligent farmers to stand up and acknowledge the responsibilities they bear in the matter of public health. He writes as well to the alarmed consumer of agricultural products, hoping to spread the knowledge of the possibilities of protective medicine — part of a concerted attempt to remove the causes of ill health, disease and, in particular, cancer. *Softcover, 368 pages. ISBN 0-911311-64-5*

The Keys to Herd Health

BY JERRY BRUNETTI

Whether dairy or beef, a healthy herd begins in such keystone concepts as biodiversity on the farm, acid/alkali balance in feedstuffs, forage quality, and more. In this accessible video, eco-consultant and livestock feed specialist Jerry Brunetti details the keynote essential for successful livestock operation. A popular speaker at eco-farming events across North America, Brunetti explains the laws of nature in terms farmers can embrace, and doles out specific steps you can utilize on your farm right away — all in a convenient video format that you can watch and review whenever you like. *VHS & DVD format.*

Holistic Veterinary Care

BY JERRY BRUNETTI & HUBERT J. KARREMAN, V.M.D.

Dr. Hubert J. Karreman, author of the compendium *Treating Dairy Cows Naturally,* is joined by renowned animal nutrition expert Jerry Brunetti to present an overview of the strategies and tools available for successful holistic herd health management. The emphasis is on natural alternatives for the treatment of common dairy cow problems, including complications in reproduction, birth and lactation. This video will provide you with a basic understanding of the power and the limitations of herbs, how to treat the whole cow, and how to build a herbal medicine kit for your farm. Drawing on actual case studies, which are examined, diagnosed, and treated using holistic protocols, this video serves as a virtual hands-on course in holistic herd health that will prove invaluable to every dairy producer, from the micro-scale family farmer to commercial-scale operations. *VHS & DVD format.*

The Other Side of the Fence — Historic Video

WITH WILLIAM A. ALBRECHT, PH.D.

Professor William A. Albrecht's enduring message preserved and presented for future generations. In this 1950s-era film, with introductory and closing remarks by Acres U.S.A. founder Charles Walters, Prof. Albrecht explains the high cost of inadequate and imbalanced soil fertility and how that "dumb animal," the cow, always knows which plant is the healthier, even though we humans don't see a difference with our eyes. A period film that is dated in style but timeless in message. Perfect for your group gathering. *VHS & DVD format, 26 minutes.*

To order call 1-800-355-5313 or order online at www.acresusa.com

Acres U.S.A. — books are just the beginning!